TOXIC HAZARDS OF RUBBER CHEMICALS

TOXIC HAZARDS
OF RUBBER CHEMICALS

A. R. NUTT

Health and Safety Adviser, Dunlop Ltd, Birmingham, UK

ELSEVIER APPLIED SCIENCE PUBLISHERS
LONDON and NEW YORK

ELSEVIER APPLIED SCIENCE PUBLISHERS LTD
Ripple Road, Barking, Essex, England

Sole Distributor in the USA and Canada
ELSEVIER SCIENCE PUBLISHING CO., INC.
52 Vanderbilt Avenue, New York, NY 10017, USA

British Library Cataloguing in Publication Data

Nutt, A. R.
Toxic hazards of rubber chemicals.
1. Rubber industry and trade—Hygienic aspects
2. Rubber industry and trade—Safety measures
I. Title
363.1'196782   RC965.R8

ISBN 0-85334-242-3

WITH 47 TABLES AND 7 ILLUSTRATIONS

© ELSEVIER APPLIED SCIENCE PUBLISHERS LTD 1984

All rights reserved. No part of this publication may be reproduced, stored in a retrieval system, or transmitted in any form or by any means, electronic, mechanical, photocopying, recording, or otherwise, without the prior written permission of the copyright owner, Elsevier Applied Science Publishers Ltd, Ripple Road, Barking, Essex, England

Printed in Great Britain by Galliard (Printers) Ltd, Great Yarmouth

Preface

Rubber manufacturing is essentially a chemical industry. Each rubber formulation contains a carefully chosen mixture of chemicals which will react together during vulcanisation to produce a new compound with properties suitable for its intended purpose. Many of the chemicals used in these formulations contain active groupings to enable them to effect these reactions, and perhaps because of this chemical activity many of them are likely to possess potential health hazards.

There is a bewildering variety of materials to be considered. Over 500 distinct chemicals are today being sold for use in conventional rubber processing, and a number of the chemical groupings involved are almost specific to the rubber industry. If these materials had not found use in the industry, most would have remained as oddities tucked away in the more obscure corners of organic chemistry text books. In the light of this it is not surprising that many of the materials have fallen outside the mainstream of investigations for potential toxicity. For example, it is revealing that less than 20 of these 500 materials have been given Threshold Limit Values on the American Conference of Governmental Industrial Hygienists lists.

In recent years a good deal of this missing knowledge has accumulated through the activities of the industry, the chemical suppliers and government agencies. This book attempts to set down the present position and in doing so to point out some areas where information is still lacking.

As will be seen, the book is divided into three parts. Part I reviews the investigations which have been carried out on various aspects of health in the industry. Chapter 1 of this concerns what must be the starting point for any consideration of health in the industry—the epidemiological investigations. Epidemiology is the study of health patterns within a population of people; a full epidemiological study can provide a measured answer to the question of whether the health of people working in the industry is being affected by their occupation. A considerable amount of work has now been carried out in the rubber industries of various countries and quite a detailed picture has emerged and has revealed some current problems.

As in other sectors of industry, the diseases which have received most attention, for reasons which are not always completely logical, are the cancers. The rubber industry cancer studies began with the bladder cancer problems in the UK, and the history is reviewed in Chapter 2. In more recent years, three groups of potential carcinogens have had particular attention focused on them—the aromatic amines, the polycyclic aromatic hydrocarbons and the nitrosamines. The studies on these chemicals are reviewed in Chapter 3. The nitrosamine situation illustrates a complication which may be more common in the rubber industry than elsewhere; because of the chemical reactions taking place during rubber processing, new chemical species may be generated which present different toxicity hazards from the raw materials themselves. Some work has now been carried out to study the volatile products of vulcanisation, but detailed knowledge here is far from complete.

Part II contains details of health effects for the common rubber chemicals. The chemicals concerned have been grouped into nine classes according to their function in rubber processing, and are covered in Chapters 4–12. For simplicity of presentation, a standard format has been used wherever possible for each chemical with the structure of the materials shown and information given on proprietary names, physical form, acute toxic effects, skin and eye irritation and chronic toxic effects. It is hoped that this part of the book will enable the user to find concise information on the health hazards of specific materials as easily as possible, and to obtain references to original work where required.

Many of those concerned with using rubber chemicals in industry now have a considerable responsibility for health and safety in these operations, and need to have an appreciation of the potential problems involved in using these chemicals. Unfortunately, many of the terms and methods used in toxicology and industrial hygiene will prove unfamiliar ground for the line manager in this position, who may well be tempted to conclude that 'health problems should be left to the experts'. This conclusion is not usually justified. Although specialist help may well be needed, the manager will very often have to decide when to ask for such investigations and advice, and what subjects should be investigated. In practice, also, problems are likely to be defined and solved most quickly and effectively if the line manager, the toxicologist, the industrial hygienist and the environmental engineer all have a working understanding of each other's field of operation. Part III of this book is therefore intended as a reference guide to the terminology and methods used in the rest of the book. It provides short explanations of the major types of effect of chemicals, describes the methods used in testing chemicals for toxicity, and reviews the atmospheric monitoring methods used in the industry. It is hoped that this will provide background information for the non-specialist who requires a working acquaintance with these subjects. A bibliography is provided for guidance on where to find more detailed information on specific aspects of individual topics.

The basic aim of all those working on toxicity problems in the rubber industry is to prevent or eliminate any damage to the health of those persons working in the industry. This needs to be a continuing process as new chemicals and new ways of using existing chemicals continue to make an appearance. The present pattern of health has been well defined by the epidemiologists; a continual updating of these investigations is necessary because of the inevitable time lag associated with epidemiology. Work on the mechanisms of interaction of rubber chemicals with the human body is still relatively young, but is expanding rapidly. Control of exposure to chemicals in the industry is becoming a routine part of normal plant management, and involvement of the industrial hygienist and the environmental engineer on these operations is now much more common than was previously the case.

All these efforts will be well justified if they result in the elimination of health problems such as those associated with the rubber industry bladder cancer episode, and enable similar potential problems to be avoided in the future.

A. R. NUTT

Contents

Preface v

Acknowledgements xi

PART I: HEALTH IN THE RUBBER INDUSTRY

1. Epidemiological Studies of the Rubber Industry . . . 3
2. Bladder Cancer in the Rubber Industry 24
3. Hazards from Aromatic Amines, Polycyclic Aromatic Hydrocarbons and Nitrosamines 29

PART II: TOXICITY OF RUBBER CHEMICALS

4. Natural and Synthetic Rubbers 53
5. Reinforcing Agents, Activators and Fillers 56
6. Curing Agents 63
7. Oils, Waxes, Resins and Plasticisers 68
8. Accelerators 72
9. Retarders 96
10. Antidegradants 99
11. Blowing Agents 115
12. Solvents 120

PART III: PHYSIOLOGICAL EFFECTS OF CHEMICALS, TOXICOLOGICAL TESTING AND ATMOSPHERIC MONITORING

13. The Effects of Chemicals on Health 133
14. Routes of Absorption of Chemicals 140
15. Toxicological Testing of Chemicals 145
16. Atmospheric Monitoring Methods 156

Appendix 176

Bibliography 178

Chemical Trade Names Index 183

Chemical Names Index 187

Subject Index 191

Acknowledgements

I am indebted to a number of colleagues who have provided information and helpful discussions on various topics included in this book. I would particularly like to thank the various members of the British Rubber Manufacturers' Association's Working Party on Toxic Hazards of Rubber Chemicals, among them Dr Guy Parkes, Dr Charles Veys and Dr Arthur Spivey. My thanks are also due to Dr Ian Carney, ICI Central Toxicology Laboratories, for helpful comments and suggestions on the material used in Chapters 13-15. I am grateful to Enid Skidmore for typing the manuscript and to Dunlop Ltd for their permission to publish the book. In this context, I should say that any views expressed or comments made are mine, and do not necessarily represent those of Dunlop Ltd.

A. R. N.

PART I
HEALTH IN THE RUBBER INDUSTRY

1

Epidemiological Studies of the Rubber Industry

There are various designs for epidemiological study. Those most used in the rubber industry have been of two kinds: 1. Retrospective follow-up studies. 2. Case/control studies.

RETROSPECTIVE FOLLOW-UP STUDIES (LONGITUDINAL STUDIES)

The most common variety of these studies is the mortality study. Here the rate of death from a specific disease is compared with the rate in a comparison population, usually the general population. Less commonly, morbidity studies are made. In these the rate of occurrence of a specific disease (which need not necessarily cause death) is compared with the rate in a comparison population. The results of these follow-up studies are normally given as standardised mortality (morbidity) ratios (SMRs). An SMR is defined as the ratio observed to expected deaths or cases of disease expressed as a percentage. The expected numbers are calculated from the information obtained from the comparison population.

The study population (cohort) may be defined in two main ways. In a census cohort, the population is defined as all the active employees in the industry or factory under study at one point in time. In an active intake cohort, it consists of all the workers entering the industry or factory over a defined time period.

Census cohorts suffer from unrepresentative distributions of length of exposure among the population, compared with the broader sample in an active intake cohort. In particular, those persons with the greatest lengths of exposure will tend to be under-represented in the type of study, since they may have found employment elsewhere, retired or died prior to the census date. In an active intake cohort, especially when studying cancer, it may be necessary to analyse only those deaths or cancers occurring after a given number of years have elapsed following first exposure. This is because occupational cancer is unlikely to occur soon after exposure, and inclusion of deaths or cancers in this early period may dilute any real effect which is present.

In follow-up studies, the incidence of a given cause of death is not infrequently less than in the general population. This effect, known as the 'healthy worker effect' is caused by the fact that the general health of people taken into industry is better than that in the general population. The effect is most apparent in the early years after employment, and tends to diminish with time.

CASE/CONTROL (OR REFERENT) STUDIES

In these studies the 'case' population is taken from persons who have had a particular disease and is compared with a 'control' population who have not. The control population is matched, for sex, age and sometimes other factors, with the case population, and usually three or more controls are chosen for every case. If a case/control study is taken among the general population (for instance, by using hospital records of a particular disease) the study will be seeking to find whether more of the cases have worked in the rubber industry than have members of the control population. Case/control studies have also been made inside the rubber industry. Here the object is to try to find if particular jobs in the industry seem to be associated with an excess of the specific disease.

STUDIES IN THE UK

The early studies of Case are discussed in the next chapter. In 1970, the British Rubber Manufacturers' Association (BRMA) in conjunction with the University of Birmingham, set up a retrospective follow-up study in the UK rubber industry, primarily to discover whether an excess of bladder cancer remained in the industry.

The BRMA/Birmingham University Study
The study covered 37 221 men entering the UK rubber industry between 1946 and 1960. Thirteen factories were included in the study, 8 of them

being tyre factories and 5 being general rubber goods (GRG). The population under study, all of whom had worked for at least 1 year in the industry, was divided into 3 cohorts, according to their date of entry into the industry.

Cohort 1: 1 January 1946–31 December 1950.
Cohort 2: 1 January 1951–31 December 1955.
Cohort 3: 1 January 1956–31 December 1960.

Initial Study

The initial study[1] traced the population to 31 December 1970 to find how many were alive by this date, and to record the causes of death for those who had died. The study was later extended to an end date of 31 December 1975[2] though the latter report analysed only deaths from cancer.

From the initial study, the figures in Table 1.1 were obtained. It can be seen that overall, less deaths occurred in this population than expected, a result largely explained by the healthy worker effect, although the deficit of deaths from heart disease may indicate that there is here some protective effect from working in an industry which generally requires considerable physical activity.

Cancer of certain sites was shown here to be in excess of expectation. Because of the long latent interval which occurs between exposure and

TABLE 1.1
INITIAL RESULTS OF THE BRMA/BIRMINGHAM UNIVERSITY STUDY[1]

Cause of death	Observed	Expected	SMR
All causes	4 167	4 721·2	88·3
Respiratory disease (non-malignant)	608	587·2	103·5
Ischaemic heart disease	1 021	1 162·1	87·9
External causes (including suicide)	260	448·2	58·0
All cancers	1 162	1 201·6	96·7
Cancer of lung	560	520·7	107·5
Cancer of stomach	165	147·7	111·7
Cancer of pancreas	33	41·7	79·2
Cancer of prostate	19	23·4	81·1
Cancer of bladder	33	30·8	107·2
Cancer of brain	25	37·8	66·1
Cancer of colon	60	61·1	98·1
Cancer of rectum	46	46·2	99·5
Cancer of larynx	11	9·4	117·1
Leukaemia	32	36·1	88·6

TABLE 1.2
CANCER INCIDENCES USING 10 YEAR LATENCY

Cause of death	Observed	Expected	SMR
Cancer of lung	303	245·3	125·5
Cancer of stomach	91	66·4	137·1
Cancer of pancreas	20	21·24	94·2
Cancer of prostate	12	6·14	195·6
Cancer of bladder	18	13·0	138·7
Cancer of brain	12	26·36	45·5
Cancer of colon	29	31·91	90·9
Cancer of rectum	16	22·14	85·8
Cancer of larynx	6	3·69	162·6
Leukaemia	18	28·2	63·8

occurrence of a cancer, the initial study also analysed deaths from cancer by studying only those members of the study population who had survived for at least 10 years from their date of first employment. This technique removes from the analysis those cases of death which occur relatively close to the point of entry into the industry, and which, for cancer, are unlikely to be caused by exposure in the rubber industry. The study termed this refinement '10 year latency'.

Using this technique the figures in Table 1.2 were produced. There is no evidence on these figures of any excess for cancers of the pancreas, brain, colon, rectum or leukaemia. The figures for prostate and larynx are too small to enable any firm conclusions to be drawn. The other cancers are worth considering further in some detail.

Bladder Cancer: Overall the population had an SMR of 138·7. When these figures were broken down over the three 5-year cohorts, however, the picture shown in Table 1.3 emerged. It can be seen from this that all the excess for this cancer is present in Cohort 1, in the time period when Nonox S was in use. Furthermore, when the figures in this cohort were split

TABLE 1.3
BLADDER CANCER—10 YEAR LATENCY

	Observed	Expected	SMR
Cohort 1	15	9·9	152·0
Cohort 2	3	3·1	96·7
Cohort 3	—	—	—

into occupational groups, the bulk of the excess was found in three groups: extruder operators (5 observed, 1·0 expected, SMR 503·2), component builders (6 observed, 3·1 expected, SMR 195·8) and moulders (4 observed, 1·0 expected, SMR 390·3). This result was almost exactly in line with anticipation. Since one of the prime reasons for setting up the study was to examine cancer of the bladder, the figures gave considerable cause for satisfaction. They provided good evidence that removal of Nonox S had essentially eliminated the bladder cancer excess and risk from the industry.

Lung Cancer: The overall SMR for lung cancer was 123·5. The lung cancer cases were also analysed into occupational groups (Table 1.4). It should be noted when attempting this type of analysis that it is often difficult to get complete details of occupational history for a retrospective study of this type, and less precise results can be expected from the smaller sub-groups being analysed. In spite of this, there is some suggestion that the lung cancer excesses are to be found in the curing, inspection and mixing areas.

This breakdown uses external comparisons to derive the expected figures. The lung cancer cases were subsequently analysed by internal comparison. For each case a 'Workogram' was constructed detailing exposure to various types of environmental contaminant over the whole of this person's occupation in the rubber industry.[3] Such workograms can be used to examine the exposure which took place over a specific period before the disease appeared, in order to find whether certain types of exposure are more common during this important 'window' in time. For each contaminant the fraction of the sub-group reported to have some exposure to

TABLE 1.4
LUNG CANCER WITH OCCUPATION

Occupation	*SMR—10 year latency*
Stores	98·6
Mixing	158·7
Latex	96·5
Extruding, calendering	119·7
Building	126·2
Curing	140·9
Inspecting	198·3
Finishing	102·4
Labouring, site workers	67·2
Maintenance	107·0

TABLE 1.5
LUNG CANCER WITH TYPE OF EXPOSURE

Contaminant	Fraction of total group exposed to this environment	
	All cases (595 cases), %	10 Year latency (368 cases), %
Vulcanised fume	45	51
Unvulcanised fume	34	37
Solvent	43	44
Talc	27	30
Powder	21	24

the contaminant was calculated (Table 1.5). This analysis suffers from the fact that it is uncontrolled—no calculation has been made of the standard rubber worker's environment. In addition, no account has been taken of age. In spite of this, the 'vulcanised fume' environment shows the greatest enhancement of association by use of the latency technique.

In an attempt to rectify these deficiencies, a case control study was carried out. Four controls were chosen for each lung cancer case, matched for age, factory, cohort and duration of service. The environmental data

TABLE 1.6
LUNG CANCER—CASE CONTROL STUDY (514 CASES)

Contaminant	Number of controls reported as exposed to this environment ($\times 0.25$)	Number of lung cancer cases reported as exposed to this environment	Ratio lung cancer/controls[a] (%)
Vulcanised fume	96.8	114	118
Unvulcanised fume	139.3	133	96
Solvent	115.3	103	89
Talc	87.5	91	104
Other powder	63.3	76	120
Mineral oil	7.5	6	80
Rubber dust	10.8	4	37
No significant exposure	206.8	201	97
Other exposure and unknown	70.8	76	107

for the 2500 cases and controls was collected (Table 1.6). It is clear from the table that no single environment stands out as a cause of these cancers. Once again, however, vulcanised fume was highlighted by this analysis, and in addition the 'other powder' contaminant was more frequently present in cases than controls.

Overall, the various analyses which have been made of these lung cancer cases suggest that they may be associated with exposure to vulcanising fume, and perhaps also with exposure in the mixing area.

Stomach Cancer: The overall SMR for stomach cancer was 137·1. The same types of analysis were made on these figures as described for lung cancer (see Tables 1.7–1.9). The analyses show that the cases of stomach cancer were in general diffusely spread across various occupations and exposures. No clear cut conclusion can be drawn from this data regarding the association of stomach cancer with types of occupation or exposure.

TABLE 1.7
STOMACH CANCER WITH OCCUPATION

Occupation	SMR—10 year latency
Stores	—
Mixing	161·4
Latex	103·0
Extruding, calendering	168·7
Building	134·6
Curing	115·1
Inspecting	149·2
Finishing	201·7
Labouring	146·1
Maintenance	125·6

TABLE 1.8
STOMACH CANCER WITH TYPE OF EXPOSURE—WORKOGRAM STUDY

Contaminant	Fraction of total group exposed to this contaminant
Vulcanised fume	40
Unvulcanised fume	39
Solvent	35
Talc	28
Powder	20

TABLE 1.9
STOMACH CANCER—CASE CONTROL STUDY (151 CASES)

Contaminant	Number of controls reported as exposed to this contaminant ($\times 0.25$)	Number of stomach cancer cases reported as exposed to this environment	Ratio stomach cancer/controls (%)
Vulcanised fume	39.5	30	75
Unvulcanised fume	44.8	46	103
Solvent	37.8	33	87
Talc	35.0	30	86
Other powder	25.8	24	93
Mineral oil	3.3	3	91
Rubber dust	2.3	4	174
No significant exposure	56.5	55	98
Other exposure and unknown	14.5	16	110

Extended Study

The extended BRMA study, reported in 1982[2] traced the original population to an end date of 31 December 1975 giving a further 5 years of analysis. It concentrated on cancer mortality, and again used the technique of excluding from the study those individuals who had died in the 10 years following their date of first employment. This extra restraint reduced the initial population from 36 695 to 33 815 men. The effects of this on death from all causes showed that the 'healthy worker effect' is largely dissipated by a 10 year period of employment (Table 1.10). The overall results for cancer mortality in the reduced population are as shown in Table 1.11. Once again there is no evidence from these figures of any excess for cancers of brain, colon or rectum, and in addition of prostate or bladder. The number of thyroid cancer cases was so small that any interpretation is

TABLE 1.10
EXTENDED BRMA/BIRMINGHAM UNIVERSITY STUDY

Deaths from all causes	Observed	Expected	SMR
Initial population	6 340	6 504	97.5
Reduced population (excluding deaths in first 10 years)	4 882	4 841	100.8

TABLE 1.11
RESULTS OF EXTENDED BRMA/BIRMINGHAM UNIVERSITY STUDY

Site of cancer	Observed	Expected	SMR
All cancers	1 359	1 221·0	111
Lung	638	517·4	123
Stomach	183	141·8	129
Prostate	30	45·4	66
Thyroid	4	2·2	182
Colon	67	73·7	90
Rectum	50	54·1	92
Oesophagus	40	31·8	126
Brain	35	41·1	85
Leukaemia	31	28·1	110
Bladder	36	43·0	83
Others	245	242·4	101

difficult. Leukaemia was very marginally in excess of expectation but the excess was not statistically significant. Three sites of cancer produced excess figures.

Lung Cancer: The excess found in the previous study was confirmed. When the results for each of the three cohorts was obtained, it could be seen that most of this excess is found in the earliest cohort. The study also split the cases according to occupation (tyre and general rubber goods) and as can be seen from the table, results were rather worse in the GRG factories (see Table 1.12). The results in Table 1.13 were obtained for the various occupational groups.

TABLE 1.12
LUNG CANCER RESULTS BY COHORT AND INDUSTRY

	Industry	Observed	Expected	SMR
Cohort 1	Tyre	348	279·8	124
	GRG	84	46·4	181
Cohort 2	Tyre	120	114·6	105
	GRG	30	25·2	119
Cohort 3	Tyre	47	43·3	109
	GRG	9	8·2	110

TABLE 1.13
LUNG CANCER WITH OCCUPATION

Occupation	Tyre			GRG		
	Observed	Expected	SMR	Observed	Expected	SMR
Stores	8	6·7	119	2	1·8	111
Mixing	42	38·7	109	19	10·5	181
Latex	15	19·8	76	3	2·1	143
Extruding, calendering	45	38·7	113	17	12·5	136
Building	199	161·8	123	31	21·0	148
Curing	68	56·1	121	20	13·2	152
Inspection	25	15·3	163	11	4·1	268
Finishing	46	32·1	143	8	4·1	195
Labouring, site workers	45	33·8	133	6	5·1	118
Maintenance	92	89·0	103	21	14·3	147

These results may indicate some variation according to type of factory. Both groups show excesses in the building, curing, inspection and finishing areas. In addition, the GRG factories show some excesses in the mixing, extruding and calendering, and maintenance operations.

Stomach Cancer: The excess found in the original study was confirmed. Unlike lung cancer, however, some of the excess was found in the tyre section of Cohort 3 (see Table 1.14). When split into occupational groups, the numbers become rather small, as shown in Table 1.15.

Perhaps the only trends which can be discerned here are the excesses in mixing (both groups) and labouring/site workers (tyre group).

TABLE 1.14
STOMACH CANCER RESULTS BY COHORT AND INDUSTRY

	Industry	Observed	Expected	SMR
Cohort 1	Tyre	96	78	123
	GRG	22	13·3	165
Cohort 2	Tyre	33	30·5	108
	GRG	8	6·7	119
Cohort 3	Tyre	22	11·2	196
	GRG	2	2·1	95

TABLE 1.15
STOMACH CANCER WITH OCCUPATION

Occupation	Tyre			GRG		
	Observed	Expected	SMR	Observed	Expected	SMR
Stores	4	1·8	222	1	0·5	200
Mixing	17	10·3	165	4	2·8	143
Latex	4	5·4	74	0	0·6	—
Extruding, calendering	15	10·4	144	5	3·4	147
Building	52	43·6	119	9	5·7	158
Curing	18	14·7	122	2	5·7	158
Inspection	7	3·9	179	2	1·2	167
Finishing	4	9·0	44	2	1·2	167
Labouring, site workers	24	10·6	226	2	1·7	118
Maintenance	20	24·2	83	6	3·9	154

Oesophageal Cancer: Unlike the earlier BRMA study, this study found a small excess of deaths from oesophageal cancer (40 observed, 31·8 expected, SMR = 126). Both tyre and GRG sectors were involved. The report raises the possibility that oesophageal and stomach cancer may share a common cause.

The Health and Safety Executive (HSE) Study

This study used a census cohort of 40 867 men aged 35 or over working in the rubber industry on 1 February 1967. Only men who had worked for at least a year were included in the study. The cohort was followed to 1 February 1977. Three sub-groups within this cohort were defined:

(A) Men who started work before 1 January 1950 in a factory using suspect antioxidants (e.g. Nonox S).
(B) Men who started work after 1 January 1950 in a factory which had previously used suspect antioxidants.
(C) Men who had worked in factories which had never, so far as could be determined, used suspect antioxidants.

The men were also classified as to their occupation on the census date of 1 February 1967. Classification of jobs on a single date such as this is however unlikely to give a true picture of occupations for each member of the population, since these change over the working lifetime.

TABLE 1.16
RESULTS OF THE HSE STUDY

Cause of death	Observed	Expected	SMR
All causes	5 773	5 912·5	98
Respiratory disease (non-malignant)	659	711·4	93
Ischaemic heart disease	1 983	2 014·6	98
Diabetes mellitus	31	37·4	83
All cancers	1 776	1 617·2	110
Cancer of lung, trachea, bronchus	822	716·5	115
Cancer of stomach	216	176·4	122
Cancer of pancreas	66	70·8	93
Cancer of prostate	55	55·8	99
Cancer of bladder	73	57·6	127
Cancer of brain	21	34·8	60
Cancer of colon	107	93·2	115
Cancer of rectum	75	70·6	106
Cancer of oesophagus	35	44·0	80
Cancer of lymphatic and haematopoietic tissue	77	89·9	86
Leukaemia	33	33·8	98

Expected rates of death from various causes were obtained from national statistics, sometimes with allowance for regional variations and social class.

The study produced the overall results shown in Table 1.16. Once again excesses were observed for deaths from lung, stomach and bladder cancer. A small excess of colon cancer was found. No excess was identified for cancers of pancreas, prostate, brain, oesophagus and lymphatic and haematopoietic tissue, including leukaemia. The similarity of these results to the BRMA results, despite the different methodology, must be regarded as important.

Bladder Cancer: The analysis of bladder cancer cases by the three sub-groups proved valuable (Table 1.17). The table shows that the excess

TABLE 1.17
RESULTS OF SUB-GROUPS

Sub-group	Observed	Expected	SMR
A	36	25	144
B	24	22·3	108
C	13	13·4	97

TABLE 1.18
TYRE AND NON-TYRE RESULTS FOR SUB-GROUP A

Sub-group A	Observed	Expected	SMR
Tyre	13	13·4	97
All non-tyre	23	11·6	198
Cable and electrical	6	1·9	314
Mouldings, motor accessories and mechanicals	13	5·5	238

bladder cancer identified was almost wholly located in sub-group A, consisting of men who could have been exposed to Nonox S or similar antioxidants.

A further breakdown of these figures showed that this excess was located in non-tyre operations, and particularly the cable and moulding, motor accessories and mechanical sectors (Table 1.18).

The failure to observe a bladder cancer excess in the tyre sector can perhaps be explained by the fact that this type of census study, taken on a day 17 years after withdrawal of Nonox S, contains only those people who had remained in the industry for this period of time, and does not include those who have died before the census date, found employment elsewhere or retired. It should also be pointed out that the absolute number of cases in these sub-groups are small, making the statistical comparison difficult to interpret.

TABLE 1.19
LUNG CANCER BY INDUSTRY

	Observed	Expected	SMR
Tyre	326	299·4	109
All non-tyre	496	465	107
Adhesives, rubber solutions	21	11·9	176
Belting, hose, rubber with asbestos	99	71·2	139
Ebonite and vulcanite	18	9·6	186

Lung Cancer: The lung cancer excess occurred in both the tyre and non-tyre sectors. Results are given in Table 1.19.

Stomach Cancer: The stomach cancer excess was also found in both tyre and non-tyre operations. Results are given in Table 1.20.

TABLE 1.20
STOMACH CANCER BY INDUSTRY

	Observed	Expected	SMR
Tyre	91	73·9	123
Non-tyre	125	117·1	107

Veys Studies[5]
Veys has made specific studies of bladder cancer in UK tyre factories. These studies strongly reinforce the conclusion drawn from the BRMA and HSE studies that the bladder cancer excess has disappeared following removal of Nonox S and equivalent antioxidants. The results are summarised in Table 1.21.

TABLE 1.21
BLADDER CANCER RESULTS IN TWO FACTORIES

	Observed bladder tumour registrations	Expected	SMR
Factory A, population (i) (2 081 men employed between 1946–1949, followed to 1970)	23	10·3	223
Factory B (3 867 men employed between 1945–1949, followed up for 20 years)	26	13·2	197
Factory A, population (ii) (2 846 men employed after 1 January 1950, followed up for varying periods between 19 and 29 years)	6	4·5	133

STUDIES IN THE USA

The major studies of the rubber industry in the USA have been made at the University of North Carolina and the Harvard School of Public Health.

University of North Carolina (UNC) Studies
Two main populations were defined in these studies. The first comprised 6678 men who were alive on 1 January 1964 and had worked for at least

TABLE 1.22
RESULTS OF THE UNC STUDY

Cause of death	Observed	Expected	SMR
All causes	1 983	2 033	98
Chronic respiratory disease	61	68·5	89
Ischaemic heart disease	870	916	95
Arteriosclerosis	89	60·1	148
Diabetes mellitus	54	37·2	145
Cancer of lung, etc.	106	115	92
Cancer of stomach	40	23·4	171
Cancer of pancreas	18	21·9	82
Cancer of prostate	53	37·8	140
Cancer of bladder	10	13·3	75
Cancer of brain, CNS	4	6·15	65
Cancer of colon	42	34·4	122
Cancer of lymphatic and haematopoietic systems	43	31·6	136
Lymphosarcoma and Hodgkins' disease	15	9·15	164
Leukaemia	17	13·5	126

10 years in a large tyre factory in Akron, Ohio, which also contained a synthetic rubber plant and a reclaim plant. This census population was followed for 9 years from 1964.[6]

Overall the results were as shown in Table 1.22. From the table it can be seen that arteriosclerosis, diabetes mellitus and cancers of the stomach, prostate, colon, lymphatic and haematopoietic systems, lymphosarcoma and leukaemia were all in excess of expectation. Unlike the British studies, the overall rate of lung cancer was not elevated.

Certain of the causes of death were further analysed using case control techniques.[7] The job history of cases was compared with that of a control group of workers not having the disease. The analysis pointed to positive associations between the various diseases and particular jobs, as indicated by Table 1.23.

A small parallel study was also made on 1339 men in a second tyre plant in Akron chosen in the same way as for the first population.[7] The two most elevated causes of death in this population were stomach cancer (SMR 119) and lymphatic and haematopoietic cancer (SMR 176).

The lymphatic and haematopoietic cancers were further analysed using a case control study.[8] Eighty-eight cases were compared with 264 controls, matched for sex, race, plant, and age at death. All 352 members of the study

TABLE 1.23
UNC CASE/CONTROL STUDY RESULTS

Disease	Job	Number of cases	Risk ratio
Stomach cancer	Compounding and mixing	5	2·0
	Extruding, tread cementing	5	2·3
Colorectal cancer	Extruding, tread cementing	7	2·2
	Maintenance	5	1·8
Respiratory cancer	Receiving and shipping	11	1·9
	Compounding and mixing	12	1·4
	Mill mixing	5	2·1
	Extruding, tread cementing	8	1·4
	Reclaim	10	2·3
Prostate cancer	Compounding and mixing	6	1·6
	Calendering	9	2·4
	Janitoring and trucking	16	3·5
Lymphatic and haematopoietic cancers	Compounding and mixing	5	1·4
	Inspection and finishing	6	2·0
	Synthetic plant	6	6·2

were categorised into 17 job groupings each of which was rated into high, medium or light exposure to solvents. This type of exposure was chosen since exposure to benzene has been shown to be leukaemogenic. Among the causes of death, only lymphatic leukaemia was shown to be associated with solvent exposure, with cases 3·3 times more likely to be exposed to solvent than controls.

A separate case control study was carried out on the prostate cancer cases within the original population.[9] Eighty-eight cases of prostate cancer were compared with 258 matched controls. Certain categories of jobs in the mixing area (batch preparation and service to batch preparation) appeared to be associated with prostate cancer, though a total of only 17 cases was involved here.

In spite of the overall lack of an excess of lung cancer, a case control study was carried out on 61 lung cancer cases and 61 matched controls.[10] This study showed no correlation between tyre building and lung cancer, but a strong correlation between curing and lung cancer—25% of the cases were exposed to curing against 15% of the controls; the average duration of exposure to curing of the cases was 16 years, against 7·5 years for the controls.

The second population studied by the UNC comprised 8418 white male

TABLE 1.24
SECOND UNC STUDY RESULTS

Cancer site	Observed	Expected
Stomach	34	27·7
Large intestine	53	45·7
Pancreas	34	27·9
Prostate	59	45·9
Bladder	21	18·1
Lymphatic and haematopoietic	52	41·9

production workers, who were followed from 1 January 1964 to 13 December 1973.[11] Certain cancers were shown to be in excess, as shown in Table 1.24.

A subsequent report[12] assigned jobs to members of the population, based on the most representative department. Positive associations were found between

(i) stomach cancer and milling (6 observed, 1·6 expected);
(ii) prostate cancer and general service (10 observed, 4·7 expected);
(iii) leukaemia and general service (6 observed, 2·4 expected).

A separate case control study of stomach cancer was carried out on cases from both the main populations defined above.[13] The job categories most strongly associated with stomach cancer were batch preparation and cure preparation. The individual jobs were also assigned according to estimated exposure to polycyclic hydrocarbons, nitrosamines, carbon black and talc. The first three contaminants showed no correlation with stomach cancer, while exposure to talc gave a positive association.

Studies at the Harvard School of Public Health

The initial study carried out by the Harvard workers defined a population of 13 570 white male workers employed at one rubber plant in Akron between 1940 and 1971. All members of the population had worked for at least 5 years in the factory. The mortality experience between January 1940 and June 1974 was studied.[14]

Overall, the number of deaths in this population was less than expected from US national statistics. The total number of deaths from cancer was also less than expected, although certain cancers did appear to be present in excess (Table 1.25). The bladder cancer excess was greatest in men who had worked at least 35 years and who died at age 75 or above. No excess risk was

TABLE 1.25
HARVARD STUDY RESULTS

	Observed	Expected	SMR
All cancers	984	1 046·4	94
Stomach	98	93·9	104·4
Bladder	48	39·5	121·5
Leukaemia	55	43	127·9

present in men who started working after 1934. The leukaemia excess was also present only among men who started work before 1935.

Subsequently, a cancer morbidity study was carried out at this plant giving correlations between the occurrence of particular cancers and job categories.[15] The excesses shown in Table 1.26 were found.

TABLE 1.26
HARVARD RESULTS—CANCERS BY JOB

Cancer	Job	Observed	Expected
Lung cancer	Tyre curing	31	14·1
	Tyre moulds	10	5·0
	Fuel celis/deicers	46	28·8
Bladder cancer	Tyre building	16	10·7
Skin cancer	Tyre building	12	1·9
Brain cancer	Tyre building	8	2·0
Lymphatic cancer	Tyre building	8	3·2
Leukaemia	Calendering	8	2·2
	Tyre curing	8	2·6
	Tyre building	12	7·5

The original study was subsequently extended[16] to an end date of 30 June 1978, and widened by including persons who had worked 2–5 years. The population was categorised as 'front processing' (compounding, mixing, milling) and 'back processing' (extrusion, calendering, production of rubber solutions and rubberised fabric). The two groups showed differences in the types of cancer found to be in excess as shown in Table 1.27. Again in this study, most of the cancer excess occurred among men who started work before 1950.

TABLE 1.27
CANCERS BY OCCUPATION

	Site	Observed	Expected
Front processing	All digestive cancers	51	34·0
	Stomach	15	7·1
	Large intestine	19	11·2
Back processing	Biliary and liver cancer	9	5·0
	Leukaemia	14	5·8

STUDIES IN FINLAND, SWITZERLAND AND SWEDEN

In Switzerland a study was made to compare the rates of death from certain diseases in a population working in the rubber industry with the equivalent rates in a population working in an explosives factory.[17] The populations were followed from 1 January 1955 to 31 December 1975. Expected numbers were derived from the general population rates. For the rubber industry, excesses were found for bladder and stomach cancer; bladder cancer was also found to be in excess in the explosives population (see Table 1.28).

In Finland, a population of 1331 male and female workers in a rubber factory were observed from 1 January 1953 to 31 December 1976. For all cancers, 21 cases were found and 18·9 expected. Two cases of bladder cancer were found and 0·3 expected.[18]

In Sweden,[19] a significant increase in oesophageal cancer was identified among vulcanisation workers (8 observed, 0·794 expected, SMR = 1008). An increase in lung cancer was also identified in this group (13 observed, 7·4 expected, SMR = 175).

TABLE 1.28
CANCER EXCESSES IN RUBBER INDUSTRY AND EXPLOSIVES FACTORIES IN SWITZERLAND

	Site of cancer	Observed	Expected
Rubber industry	Bladder	4	1·1
	Stomach	8	4·6
Explosives factory	Bladder	5	2·6

CONCLUSIONS ON EPIDEMIOLOGY

Although there are detailed differences between the various studies which have been made, there are clearly some common trends. Death from all causes in the industry is no higher than would be expected in the general population, but certain diseases have been established to occur more frequently than expected, although the excesses are relatively small.

Stomach cancer is found in excess in both the UK and the USA. The detailed analysis of the US studies indicates that this cancer may be associated with work at the front end of the rubber manufacturing process, where dust is likely to be a problem, though the UK studies are less clear on this association.

Lung cancer excesses have also been found in the UK and some of the US studies. Analysis of the UK figures provides evidence for a connection between this lung cancer excess and exposure to hot rubber fumes. Groups of workers who would be exposed to this type of fume have also been shown to have an excess rate of lung cancer in the US studies.

Oesophageal cancer, though less common than the above two cancers, has also been shown to be in excess in some of the UK and US studies, and in one study in Sweden.

REFERENCES

1. *BRMA Health Research Project Report*, British Rubber Manufacturers' Association, Birmingham, January 1976.
2. Parkes, H. G., Veys, C. A., Waterhouse, J. A. H. and Peters, A., Cancer mortality in the British rubber industry, *Brit. J. Indust. Med.* (1982), **39**, 209–20.
3. Veys, C. A., *North Staffs Medical Inst. J.* (1979), **XI**, 24–36.
4. Baxter, P. J. and Werner, J. B., *Mortality in the British Rubber Industries 1967–76*, HMSO, London, 1980.
5. Veys, C. A., *Conference on Health and Safety in the Plastics and Rubber Industries, University of Warwick, 1980*, Plastics and Rubber Institute, London, 1980, pp. 13.1–13.9.
6. McMichael, A. J., Spirtas, R. and Kupper, L. L., *J. Occ. Medicine* (1974), **16**, 458–64.
7. McMichael, A. J., Spirtas, R., Gamble, J. F. and Tousey, P. M., *J. Occ. Medicine* (1976), **18**, 178–85.
8. McMichael, A. J., Spirtas, R., Kupper, L. L. and Gamble, J. F., *J. Occ. Medicine* (1975), **17**, 234–9.
9. Goldsmith, D. F., Smith, A. H. and McMichael, A. J., *J. Occ. Medicine* (1980), **22**, 533–41.

10. McMichael, A. J., Andjelkovic, D. A. and Tyroler, H. A., Cancer mortality among rubber workers: an epidemiologic study, *Annals NY Acad. Sci.* (1976), **271**, 125–37.
11. Andjelkovich, D., Taulbee, J. and Symons, M., *J. Occ. Medicine* (1976), **18**, 387–94.
12. Andjelkovich, D., Taulbee, J., Symons, M. and Williams, T., *J. Occ. Medicine* (1977), **19**, 397–405.
13. Blum, S., Arp, E. W., Smith, A. H. and Tyroler, H. A., Stomach cancer among rubber workers—an epidemiological investigation, *Dust and Diseases*, eds R. Lemen and J. M. Dement, Pathotox, Illinois, 1979, pp. 325–34.
14. Monson, R. R. and Nakano, K. K., *Amer. J. Epidem.* (1976), **103**, 284–303.
15. Monson, R. R. and Fine, L. J., *J. Nat. Cancer Inst.* (1978), **61**, 1047–53.
16. Delzell, I. S. and Monson, R. R., *J. Occ. Medicine* (1982), **7**, 539–45.
17. Bovet, P. and Lob, M., *Schweiz med. Wochenschr* (1980), **110**, 1277–87.
18. Kilpikari, I., Mortality among male rubber workers in Finland, *Arch. environ. Health* (1982), **37**(5), 295–9.
19. Norell, S., Ahlbom, A., Lipping, H. and Österblom, L., Oesophageal cancer and vulcanisation workers, *Lancet* (1983), 462–3.

2

Bladder Cancer in the Rubber Industry

The UK rubber industry bladder cancer episode, which was first brought to light by the work of Case in 1949,[8] has had very important consequences for the rubber industry throughout the world. Firstly, the number of people whose health was affected was relatively large—certainly more than 100 and perhaps several hundred. Secondly, the long induction period between exposure and appearance of the cancer means that cases are still appearing more than 30 years after removal of the antioxidants which caused the problem.

Case's study showed how this virtual epidemic of cancers could be clearly identified by careful epidemiological work. Conversely, it demonstrated how even a large effect on health could go unnoticed in the absence of such studies. The methods used by Case in both the rubber industry and the chemical industry have formed the basis of modern epidemiological studies.

The bladder cancer episode has made the rubber industry acutely aware of the health problems which can be caused by chemicals, and it is now perhaps the best studied sector of general industry in this respect.

The early history and subsequent developments concerning bladder cancer in the rubber industry are worth reviewing in some detail. The problem originated in Germany, the birthplace of the synthetic dyestuffs industry, where aromatic amines were the most important starting material. As far back as 1895, Rehn, a German surgeon, described three cases of bladder cancer in dyestuffs works.[1] By 1912, the aromatic amines β-naphthylamine and benzidine had been clearly implicated as causes of occupational bladder cancer in dyestuffs manufacture in Germany and Switzerland.[2,3] In 1921, an ILO study of the position in these countries concluded that benzidine and β-naphthylamine were the active agents.[4] The UK dyestuffs industry took longer to recognise this problem although

Ross in 1918 referred to 14 cases of villous growth of the bladder in workers handling aniline dyes,[5] and Wignall in 1929 reported on the incidence of bladder cancer in the dyestuffs industry.[6] However, it was not until after the Second World War that formal studies were put in hand to determine whether dyestuffs manufactured in the UK were causing a bladder cancer problem.[7] The Research Associate engaged to undertake this work was Dr R. A. M. Case. Most of the chemical manufacture which was the subject of the study took place in Lancashire and Yorkshire and Case decided that he would also study the incidence of bladder cancer in a control area, which did not contain this type of chemical industry. For his control area he chose Birmingham County Borough, which was where he had qualified in medicine, and which had little chemical industry. On commencement of the study of the cases of bladder cancer contained in the Birmingham area records, he was surprised to find that no fewer than 40 of these cases had worked at a single rubber factory in the area. This number was considerably more than the expectation of bladder cancer for the whole of the Birmingham County Borough area. Case therefore undertook a formal study of bladder cancer in the rubber industry. For the national industry he estimated that 15·9 bladder cancer deaths would occur during the period 1936–1951 whereas he found 26. For the Birmingham area during the period 1936–1950, he estimated 4·0 cases of bladder cancer and found 22. This study was undertaken in 1949, and published in 1954.[8]

It was clear that the rubber industry had a considerable bladder cancer problem; what could be causing it? The dyestuffs studies carried out on the continent had revealed β-naphthylamine and benzidine as major causes of occupational bladder cancer. Benzidine had been used in the rubber industry as a hardener, but to a very limited extent. β-Naphthylamine was not itself used in the industry, but one of the most important antioxidants, Nonox S, which had been in use since 1928, was based on β-naphthylamine. Nonox S (an ICI product) was made by reacting acetaldehyde (paraldehyde) with a mixture of α- and β-naphthylamine. Originally (in 1928) the starting mixture of naphthylamines consisted of 50% of the β-isomer,[9] though by the time production was properly under way a mixture of 3 parts of α-naphthylamine to 1 part of β-naphthylamine was used.[10] In early 1948, increasing suspicions concerning the number of bladder cancers occurring during manufacture of the antioxidant resulted in the proportion of β-naphthylamine being halved, so that a ratio of 6 parts of α-naphthylamine to 1 of β-naphthylamine was used.[10] Nonox S did in fact contain a certain amount of residual naphthylamines. The exact quantities probably varied over the period of production, and from batch to batch, and evidence is lacking as to these variations. However, figures of 2·5 % total naphthylamine and 0·25 % β-naphthylamine have been given as representative.[11]

Nonox S was withdrawn from use in mid-1949. Subsequent epidemiological study has shown that among people employed since 1949, no excess bladder cancer has been found in the rubber industry, and it can be quite confidently inferred that Nonox S and similar antioxidants were the major cause of occupational bladder cancer in the UK rubber industry.

Case's work in the chemical industry showed that the average latent period between exposure and appearance of a bladder cancer was 16 years, although individual cases could vary from a few years to over 40.[8] Although Nonox S was withdrawn in 1949, many cases of bladder cancer have subsequently occurred. There is a natural background of bladder cancer in the general population, and it is impossible to say with certainty whether an individual case is attributable to industrial exposure or not. Nothing in the clinical appearance of the tumour distinguishes the occupational case from that which is due to natural causes. Bladder cancer in the general population does, however, become more common with increasing age, and the occurrence of bladder cancer in a younger person who is at risk from his exposure in industry is strong reason to presume an occupational causation.

In 1957, the Rubber Manufacturers' Employers Association set up a cytodiagnostic unit in Birmingham to aid in the early identification of bladder papillomas. This unit still (in 1983) carries out every year some 18 000 screening tests on urine samples provided by people who worked in the rubber industry prior to 1949.

In 1970, a test case was brought in the British High Court by Cassidy and Wright, two rubber workers who had contracted bladder cancer and who claimed damages from ICI and from Dunlop. Judgement was given for the plaintiffs,[12] and upheld on appeal. It was ruled that ICI should have been aware, before 1949, of the possibility that traces of β-naphthylamine remaining in Nonox S could have posed a health risk. Since ICI had not disclosed the composition of Nonox S, the court ruled that Dunlop could not be expected to possess this knowledge, and were therefore not guilty of negligence in using the material. The case established that a supplier is liable for the safety of his product not only during its manufacture, but during any use of it which he can reasonably expect.

Other countries appear to have been less affected by the bladder cancer problem than the UK, perhaps because antioxidants of the Nonox S type were more popular in the UK than elsewhere. In the USA, one study showed a deficit of bladder cancer in one plant, and a small excess in two. This excess was greatest among those first employed before 1935 and among those who had worked in the industry for more than 35 years.[13,14]

A case control study carried out in 5 USA tyre and rubber companies compared the work patterns of 220 workers who had died from bladder cancer with 440 matched controls who had died from other causes. It was shown that the workers who had developed bladder cancer were more likely than controls to have worked in three specific areas: milling (1·9), calender operation (2·2) and final inspection (1·5).[15]

A case control study carried out in Boston compared 470 patients with cancers of the lower urinary tract with 500 controls from the general population. 51 of the bladder cancer patients had had an occupation in the rubber industry compared with an expected number (derived from the controls) of 34·6. By restricting the analysis to main occupations, 19 of the cases were rubber workers with 11·6 expected.[16] A similar study was carried out in Canada where 5 of the cases and only 1 control reported having been rubber workers.[17]

One factor which may have affected the bladder cancer statistics in the USA was the use there of an antioxidant made by condensation of acetone and 4-aminodiphenyl (Santoflex B). 4-Aminodiphenyl is a more potent bladder carcinogen than β-naphthylamine, and if Santoflex B contained traces of this amine, it could have had similar effects to Nonox S. Santoflex B was withdrawn in 1953.

Other European evidence of rubber industry bladder cancer is somewhat scanty. A Swiss study found 4 cases in an industry population where 1·1 was expected.[18] Studies in Finland which compared 180 cases of bladder cancer with 180 age–sex matched controls found that 2 of the cases had their predominant occupation in the rubber industry compared with none in the controls.[19]

As described in the last chapter, the major studies in the UK and USA have found that no excess of bladder cancer is now occurring in the industry. The lessons which had been learnt from this unfortunate episode must now be applied to the other problems which have been identified in the industry.

REFERENCES

1. Rehn, L., *Arch. Klin. Chir.* (1895), **50**, 588.
2. Rehn, L., Uber Blasenerkrankungen bei Anilinarbeitern, *Verh. dtsch. Ges. Chir.* (1906), **35**, 313–16.

3. Leuenberger, S. G., Die unter dem Einfluss der synthetischen Farbenindustrie beobachtete Geschwülstentwicklung, *Beitr. Klin. Chir.* (1912), **80**, 208–316.
4. *Studies and Reports*, International Labour Office, 1921, Series F, No. 1, p. 6.
5. Ross, H. C., Occupational Cancer, *J. Cancer Research* (1918), **3**, 321–56.
6. Wignall, T. H., Incidence of disease of the bladder in workers in certain chemicals, *Brit. Med. J.* (1929), **2**, 281–93.
7. *Research Project on Industrial Bladder Papilloma*, Dyestuffs Group of the Association of British Chemical Manufacturers, 1947.
8. Case, R. A. M. and Hosker, M. E., Tumour of the urinary bladder as an occupational disease in the rubber industry in England and Wales, *Brit. J. Preventative and Social Medicine* (1954), **8**, 39–50.
9. Cronshaw, C. J. J., Naunton, W. J. S. and British Dyestuffs Corp., *British Patent 280661*, 3 September 1926–24, November 1927.
10. *Cassidy and Wright versus ICI and Dunlop*, High Court of Justice, Queen's Bench Division, 9 November 1970, Evidence, Vol. 12.
11. Munn, A., Bladder cancer and carcinogenic impurities in rubber additives, *Rubber Industry* (February 1974), pp. 19–20.
12. *The Times*, 2 November 1972, Law Report for 1 November: Cassidy versus ICI and Dunlop; Wright versus ICI and Dunlop.
13. Monson, R. R. and Nakano, K. K., *Amer. J. of Epidem.* (1976), **103**, 284–96.
14. Monson, R. R. and Fine, L. J., *J. of the National Cancer Inst.* (1978), **61**, 1047–53.
15. Checkoway, H., Smith, A. H., McMichael, A. J., Jones, F. S., Monsoon, R. R. and Tyroler, H. A., A case control study of bladder cancer in the United States rubber and tyre industry. *British. J. of Industrial Medicine* (1981), **38**, 240–6.
16. Cole, P., Monson, R. R., Haning, H. and Friedell, G. H., *New England J. Medicine* (1971), **284**, 129–34.
17. Howe, G. R., *et al.*, *J. National Cancer Inst.* (1980), **64**, 701–13.
18. Bovet, P. and Lob, M., Mortality from malignant tumours in rubber workers in Switzerland: Epidemiological study 1955–1975, *Schweiz med. Wochenshr.* (1980), **110**, 1277–87.
19. Tola, S., Tenho, M., Korkala, M.-L. and Järvinen, K. E., Cancer of the urinary bladder in Finland—Association with occupation, *Int. Arch. occup. environ. Health* (1980), **46**, 43–51.

3

Hazards from Aromatic Amines, Polycyclic Aromatic Hydrocarbons and Nitrosamines

Certain chemical groupings have been the subject of detailed study in the rubber industry and the findings to date on three of these areas are reviewed here.

AROMATIC AMINES

β-Naphthylamine

β-Naphthylamine has been present in a number of the raw materials used by the industry.

Aldol α/β-Naphthylamine (Nonox S)

As far as effects on health are concerned, the most important of the β-naphthylamine-containing raw materials was the aldol naphthylamine condensation product represented in the UK by Nonox S, produced over the period 1927 to June 1949. This antioxidant was the condensation product in N hydrochloric acid solution of a mixture of α- and β-naphthylamines with excess acetaldehyde (paraldehyde). In this condensation some of the acetaldehyde reacts directly with the amines, and some reacts first to give acetaldol which then condenses with the amine:

$$\text{Naphthyl-NH}_2 + CH_3CHO \xrightarrow{N.HCl} \text{Naphthyl-N=CH-CH}_3 + H_2O$$

$$CH_3CHO + CH_3CHO \longrightarrow CH_3-\underset{\underset{OH}{|}}{CH}-CH_2CHO \xrightarrow{\text{Naphthyl-NH}_2}$$

$$\text{Naphthyl-N=CH-CH}_2-\underset{\underset{OH}{|}}{CH}-CH_3 + H_2O$$

In practice, the reaction probably goes further to give products with structures similar to the following:[1]

$$\text{Naphthyl-NH-CH(CH}_3\text{)-CH=CH-[Naphthyl-NH-CH(CH}_3\text{)-CH=CH-]}_2\text{-Naphthyl-NH-CH(CH}_3\text{)-CH}_2\text{-CHO}$$

The mixture of products was precipitated in the form of a fine powder by neutralising the solution with weak caustic soda. The precipitate was then washed and dried in ovens.

When the process was first described in 1928, the mixture of naphthylamines contained 50% α- and 50% β-naphthylamine, although this was changed to 66% α and 33% β as soon as production was properly under way. In 1948, because of suspicions about the number of bladder cancers occurring in production, the ratio was dropped to 83% α and 17% β.

Some free β-naphthylamine remained in the product and the concentration of this must have varied during the production period. The most often quoted figure for naphthylamines in Nonox S is 2·5% for total naphthylamines and 0·25% (2500 ppm) for β-naphthylamine, and this may represent the product towards the latter stages of production.[2] It is possible that greater amounts of free β-naphthylamine were present in earlier supplies of Nonox S.

Little evidence exists as to the concentrations of β-naphthylamine which were present in air during rubber processing using Nonox S. In 1949, Strafford[3] measured amine concentrations at the open door of a Banbury mixer containing a masterbatch of 2 lb of smoked sheet and 2 lb of Nonox S at 150 °C and found 290 mg m^{-3} total amines and 15 000 μg m^{-3} β-naphthylamine. Although this masterbatch contained no carbon black (the presence of which has been shown to reduce the quantity of β-naphthylamine volatilised), these figures suggest that working atmospheres around curing processes where Nonox S was used may perhaps have contained β-naphthylamine in the 1–1000 μg m^{-3} range. In this context it has been noted that heavy smokers (40 or more cigarettes per day) are exposed to about 1 μg of β-naphthylamine per day and have a two- to threefold increase of bladder cancer mortality.[4] Although this would suggest that the minimum effective amount of β-naphthylamine in man is very low, the possible presence of other bladder carcinogens in cigarette smoke or the possibility of other mechanisms for bladder cancer induction in cigarette smokers has not been excluded. A man doing a moderate level of physical work in an atmosphere containing 10 μg m^{-3} β-naphthylamine vapour would inhale 100 μg β-naphthylamine per day.

Williams[5] has reported measurements of amines at the door of the vacuum drying ovens used for production of Nonox S where a concentration of 10 000–15 000 μg m^{-3} total amines was found. A high proportion of people working on these drying ovens subsequently contracted bladder cancer.

Handling of Nonox S itself in the weighing and mixing operations would have involved exposure to dust containing β-naphthylamine. Williams gave figures for dust concentrations during de-traying and grinding of dried Nonox S, these rising to 50 mg m^{-3} (50 000 μg m^{-3}) total amines on occasions.[5]

Aldol α-Naphthylamine

Aldol α-naphthylamine, made by condensing acetaldehyde with α-naphthylamine, has also been widely used in rubber processing. It was made in the USA under the name Agerite Resin, in France under the name Antioxygene RA, and in Germany under the names Antioxidant AH and AP. Although these materials were manufactured from α-naphthylamine, some β-naphthylamine impurity is likely to have been present in the starting material. In Germany where the problems of β-naphthylamine were recognised relatively early, the α-naphthylamine was made by a process involving a relatively high degree of purification so that commercial α-naphthylamine contained only small traces of β-naphthylamine. In the UK and USA, however, commercial α-naphthylamine was less pure, originally containing 10% of β-naphthylamine which was reduced to about 4% in later years.

Some β-naphthylamine remained in the aldol α-naphthylamine antioxidants produced from this α-naphthylamine. In Antioxidant AP the concentrations have been given by Bayer as probably less than 300 ppm between 1937 and 1966, less than 30 ppm between 1966 and 1968, and less than 3–5 ppm from 1968 until production ceased. Other aldol α-naphthylamines may, however, have contained more β-naphthylamine.

Strafford reported measurements of amines by the open door of a Banbury mixer containing 2 lb of smoked sheet and 2 lb Agerite Resin at 150 °C and found 610 mg m^{-3} total naphthylamines but no detectable β-naphthylamine.[3]

Phenyl β-Naphthylamine (PBN)

This important antioxidant was introduced in this country in about 1928. It was always manufactured by condensation of β-naphthol with aniline

In spite of the fact that β-naphthylamine was not a starting material for production of PBN, the finished product contained traces of this material. Later investigations showed that some β-naphthylamine was generated during the condensation reaction, and would therefore be present in PBN even when the raw materials used were free from β-naphthylamine impurity.

In the 1960s the β-naphthylamine content of various commercial varieties of PBN was in the range 30–50 ppm, and it is likely that this concentration was also present during the early part of the manufacturing period.

ICI reported some measurements of atmospheric concentrations of β-naphthylamine during factory processing of rubber containing 1 % PBN in 1965.[6] Measurements were taken

1. over the front roll of the sheeting mill beneath an internal mixer;
2. near the feed nip of a calender;
3. over belting as it left a belt curing press (rubber cooling from 140 °C).

At none of these positions was β-naphthylamine detected, by a technique which had a limit of sensitivity of 0·7 μg m^{-3}.

In 1970 ICI introduced a modified version of PBN in which a scavenger was used during the condensation process to reduce the free β-naphthylamine content of the finished product. This version of PBN (currently manufactured by Anchor) and certain other commercial varieties of PBN still contain very small traces of β-naphthylamine, though the amount is less than 1 ppm.

In 1975 it was discovered that when PBN was ingested by some humans, a small proportion was dephenylated within the body to produce β-naphthylamine which could then be detected in the urine.[7] Only minute quantities of β-naphthylamine are produced in this manner, a maximum of 3 μg from a single oral intake of 10 mg PBN being found in one series of tests on volunteers, and 10 μg from a single oral intake of 30 mg PBN in a second test (i.e. about 10^{-4} of the PBN dose). However, these amounts were greater than the amounts of free β-naphthylamine in the PBN, showing that they were the result of metabolic dephenylation. The amount of β-naphthylamine produced seems to depend on individual variations in metabolism. As well as these controlled tests, measurements were made during normal working with PBN. Operators tipping bags of PBN into a chute were observed to produce 3–6 μg β-naphthylamine in the urine during the 24 h after the work.

β-Naphthylamine requires metabolic activation in the body in order to cause cancers, the derivatives which act directly on the cells being hydroxylated species such as 2-naphthylhydroxylamine. This hydroxylated derivative was not detected in the urine, in subsequent tests at ICI, although the detection procedures are difficult because of instability of the hydroxylated compounds.[8] It may be that dephenylation occurs at a relatively late stage in the body, and that little or no hydroxylation of the β-naphthylamine then takes place.

There is considerable epidemiological evidence that rubber workers exposed to PBN during the 1949–1970 period did not suffer from an excess rate of bladder cancer (see Chapter 2) so that it seems likely that neither the 30–50 ppm β-naphthylamine contained in this antioxidant nor the amounts produced by metabolic dephenylation were sufficient to cause cancer in practice.

Di-β-naphthyl-p-phenylenediamine (DNPPD)
This antioxidant has been used particularly in latex operations. The main commercial products used in the UK, some of which have now been discontinued, were Agerite White, Santowhite CI, Nonox CI and Antioxidant DNP. It was made by condensation of p-phenylenediamine and β-naphthol. Until the mid-1970s the β-naphthylamine content was in the region of 100–200 ppm, when the concentration was generally reduced to about 50 ppm.

In 1965 ICI reported on the atmospheric concentrations of β-naphthylamine during factory processing of rubber containing 1% of DNPPD.[6] Measurements were made at the same positions as reported for PBN, and again no β-naphthylamine could be detected, down to the limit of the test of

0·7 µg m^{-3}. Tests by Hodge[34] in the USA have suggested that no metabolic conversion to β-naphthylamine occurs with this compound.

4-Aminodiphenyl
Acetone 4-Aminodiphenyl Condensation Products
4-Aminodiphenyl was used as a raw material in the manufacture of these rubber antioxidants in the USA between 1935 and 1955 (Santoflex B, Santoflex BX, etc.). The main product of the reaction was 2,2,4-trimethyl-6-phenyl-1,2-dihydroquinoline, but it is likely that traces of unreacted 4-aminodiphenyl remained in the antioxidants. 4-Aminodiphenyl was discovered to be a potent bladder carcinogen in animal tests in 1952,[9] and the antioxidants were withdrawn shortly after this.

Acetone-Diphenylamine Condensation Products
The diphenylamine used to make these widely used antioxidants contains small traces of 4-aminodiphenyl of the order of a few ppm.[2] However, it seems likely that these traces react with the acetone during the condensation reaction, since separate tests carried out by ICI and Uniroyal have found that no detectable 4-aminodiphenyl can be found in the finished product, down to the limit of detection of 1 ppm.

Benzidine
It is sometimes claimed that benzidine was used in the rubber industry as a hardener, and use of benzidine as an accelerator was certainly investigated in the UK between 1940 and 1945. It is unlikely, however, that benzidine was regularly used for production in other than very small quantities.

Epilogue
The problems with these carcinogenic aromatic amines are now historical. All plants producing β-naphthylamine were closed in the UK by 1952. The last plant producing benzidine in the UK closed in 1962. In other countries, however, production may have continued until later years. β-Naphthylamine was still being produced in the USA in 1973[10] and in Italy in 1977.

Production, importation or use of β-naphthylamine, benzidine, 4-aminodiphenyl and 4-nitrodiphenyl were prohibited in the UK by the 1967 Carcinogenic Substances Regulations. Materials with more than 1 % of these amines were also prohibited by the Regulations. Manufacture or use of α-naphthylamine, o-tolidine, dianisidine and dichlorobenzidine were controlled by these Regulations.

POLYCYCLIC AROMATIC HYDROCARBONS (PAHs)

PAHs are a group of chemicals consisting of fused benzene rings. They are present in crude oil and are also generated by burning organic materials. Some PAHs are strong carcinogens, perhaps the best known being benzo[a]pyrene.

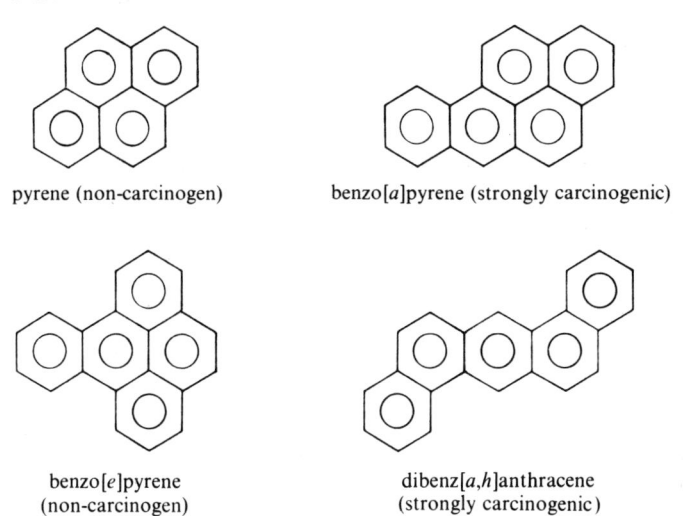

pyrene (non-carcinogen)

benzo[a]pyrene (strongly carcinogenic)

benzo[e]pyrene
(non-carcinogen)

dibenz[a,h]anthracene
(strongly carcinogenic)

PAHs occur in two of the major materials used in the rubber industry—the aromatic oils and carbon black.

Aromatic Oils

These are used in very large quantities by the industry, both as components of the oil-extended polymers supplied to the industry, and as process oils directly used by the industry. A typical car tyre tread formulation may now contain a total of 20% of aromatic oil, and total UK rubber industry consumption of aromatic oils is probably in the region of 20 000 tons per year.

These aromatic oils are produced by the oil industry as residues from the solvent refining process used on many lubricating and cutting oils for technical and health reasons. They contain large proportions of PAHs and significant concentrations of known carcinogens such as benzo[a]pyrene. Table 3.1 shows results obtained on three commercial supplies of these oils.[11] By contrast with these results a typical solvent refined oil will contain less than 1·0 ppm of benzo[a]pyrene.

TABLE 3.1
PAHs IN AROMATIC OILS

	Aromatic oil A	Aromatic oil B	Aromatic oil C
Total PAH content (%wt by DMSO extraction method):	25·9	32·7	18·2
Individual PAH contents (ppm by wt)			
Fluoranthene	11·0	0·6	1·4
Pyrene	25·6	6·4	6·9
Benz[a]fluorene	0·9	ND	0·7
Benz[a]anthracene	34·2	9·4	5·8
Chrysene	395·3	193	63·4
Benzo[b]fluoranthene	72·9	43·2	11·7
Benzo[e]pyrene	113·2	97·9	27·7
Benzo[a]pyrene	13·4	11·5	4·0
Dibenz[a,j]anthracene	4·6	5·6	2·8
Dibenz[a,h]anthracene	5·7	3·5	1·4
Indeno[1,2,3-c,d]pyrene	6·2	1·7	1·4
Benzo[g,h,i]perylene	17·9	21·1	15·3
Anthanthrene	6·6	9·8	12·3
Totals	707·5	403·7	154·8
	0·707%	0·404%	0·155%

When tested by painting on to animal skin, the oils have proved to be potent carcinogens. One test carried out by an oil supplier, using high quantities of oil on mouse skin, has given positive results for the production of skin cancer in 96% of the mice, under conditions which would produce a level of approximately 0–4% skin cancer when using a solvent refined oil. Tests such as this have resulted in warning statements being included in manufacturers' safety data sheets.[1,2]

Fortunately, the use of these oils in the rubber industry does not normally involve direct skin contact, since it is common for the oil to be piped directly from bulk storage tanks into the Banbury mixer. Even in small scale operations, the oils are normally dispensed from drums to hand held containers, and added to the mix without skin contact occurring. However, there is considerable contact with uncured rubber containing significant quantities (20%) of these oils, and this may quite often involve solvents such as petroleum naphtha which are used to wipe the surface of the rubber to increase tack. It has been suggested that such solvent use may facilitate the transfer of PAHs from the rubber to the lower layers of the skin.

The initial effects of such exposure could be expected to be similar to those seen in oilfield workers exposed to the heavier varieties of crude oil.[13] The typical lesions seen in oilfield workers are keratoses, which are relatively small, heaped-up, scaling, brown plaques on the skin, some of which may be fissured and may itch. These keratoses may progress after a variable time (which may be years) to skin cancer with ulceration. These changes are more likely in individuals with poor melamin-secreting mechanisms (blondes) and are usually due to the combined effects of sunshine and contact with oil.

Since skin cancer can normally be successfully treated, its prevalence cannot be assessed by a study of death certificates. Preliminary searches using the records of cancer registration in the UK have, however, indicated that few cases of skin cancer have occurred in the industry and that if there is a problem from this source, it is likely to be a small one. On the other hand a single report from the USA, studying hospital tumour registry records in the Akron area found 12 cases of skin cancer in a group of men with 5 or more years' experience of tyre building, when only 1·9 was expected.[14]

Carbon Black

Carbon black has also been shown to contain PAHs, and these can be extracted in the laboratory by prolonged heating with solvents.

In a comprehensive review of the environmental health aspects of carbon black, Rivin and Smith[15] gave typical concentrations of PAHs in some grades of furnace black, as shown in Table 3.2. Using this data, it is of interest to calculate the amounts of PAHs which would be present from the oils and carbon black in a practical compound.

Taking the oil to contain 700 ppm of the 13 PAHs between fluoranthene and anthanthrene given in the previous table, and 13 ppm of benzo[a]pyrene, and the carbon black 500 ppm of the 10 PAHs shown in Table 3.2, and 5 ppm of benzo[a]pyrene, a typical car tyre tread formulation might contain the concentrations of PAH shown in Table 3.3.

Carbon black has a strong surface adsorptive power for materials such as PAHs so that only small amounts of these are released from the carbon black surface by biological fluids. Buddingh showed that small amounts of benzo[a]pyrene (B[a]P) were released from furnace blacks by elution with human plasma, swine serum, swine lung homogenate and swine lung washings,[16] although these amounts were less than 0·005% of the B[a]P present on the carbon black as determined by extensive Soxhlet extraction using toluene.

TABLE 3.2
PAHs IN VARIOUS CARBON BLACKS

	\multicolumn{9}{c}{Furnace black type, ASTM designation}
	N472	N375	N339	N326	N330	N351	N660	N762	LCF4
Total extract (ppm):	400	2 100	1 200	250	290	1 300	310	800	700
Solvent (hours):	Toluene (48)	Toluene (48)	Benzene (250)	Benzene (250)	Benzene (250)	Methylene chloride (4), toluene (48)	Benzene (250)	Methylene chloride (4), toluene (48)	Toluene (48)
Individual PAHs (ppm)									
Fluoranthene	1	59	52	9	10	36	12	32	32
Pyrene	0·5	400	206	58	48	195	52	308	220
Benzo[g,h,i]fluoranthene	—	—	94	16	20	—	14	—	—
Cyclopenta[c,d]pyrene	<0·2	116	78	<0·8	20	81	13	—	5
Benzo[a]pyrene	<0·05	6·5	32	1	3	5	8	8	10
Benzo[e]pyrene	—	—	—	—	—	—	—	—	12
Indeno[1,2,3-c,d]pyrene	<0·05	13	35	1	0·3	9	7	9	4
Benzo[g,h,i]perylene	<0·05	146	163	16	25	95	41	87	120
Anthanthrene	<0·05	33	42	—	2	28	7	30	36
Coronene	0·08	219	140	3	11	92	27	80	180

TABLE 3.3
PAHs IN A RUBBER FORMULATION

	Parts by weight	% Total mix	Total of standard PAHs (ppm)	Benzo[a]-pyrene (ppm)
Oil extended SBR (including 37·5% of aromatic oil)	137·5	58·2 (including 15·9 oil)	111	2·0
Aromatic process oil	10·0	4·2	29·4	0·55
Approved HAF carbon black	75·0	31·7	158	1·6
Zinc oxide	5·0	2·1		
Stearic acid	1·0	0·4		
CBS	1·5	0·6		
IPPD	2·5	1·1		
Sulphur	1·8	0·8		
Paraffin wax	2·0	0·9		
	236·3	100·0		

Animal tests have shown that whereas the laboratory solvent extract is carcinogenic, carbon black itself is not.[17] Presumably, the amounts of material released from the carbon surface in the animal body are too small to give rise to cancers. Epidemiological studies on workers in carbon black production have shown no excess for cancers at any site[18] (see section in Part II on carbon black, pp. 56–8).

Although direct contact with the PAHs present in aromatic oils and carbon black does not seem to present a problem it is possible that these PAHs could be released during heating of the rubber compound, and exposure could then occur via inhalation. Work on this subject was originally carried out at Dunlop from 1972 onwards.[19] Samples of air were taken at various sites at Fort Dunlop on a daily basis over an 18 month period, and the amount of benzo[a]pyrene present compared with the amount found on that day in the outside air. The outside air samples showed an annual variation in the concentration of benzo[a]pyrene, with larger amounts present in winter than summer (Fig. 3.1). The variation arises from the burning of fossil fuels in domestic and industrial heating plants. No excess of benzo[a]pyrene could be identified in the factory samples (Table 3.4). The Dunlop figures were confirmed by Williams in the USA, who found the benzo[a]pyrene concentrations in tyre factories[20] as shown in Table 3.5.

Fig. 3.1. Benzo[a]pyrene levels in Birmingham air.

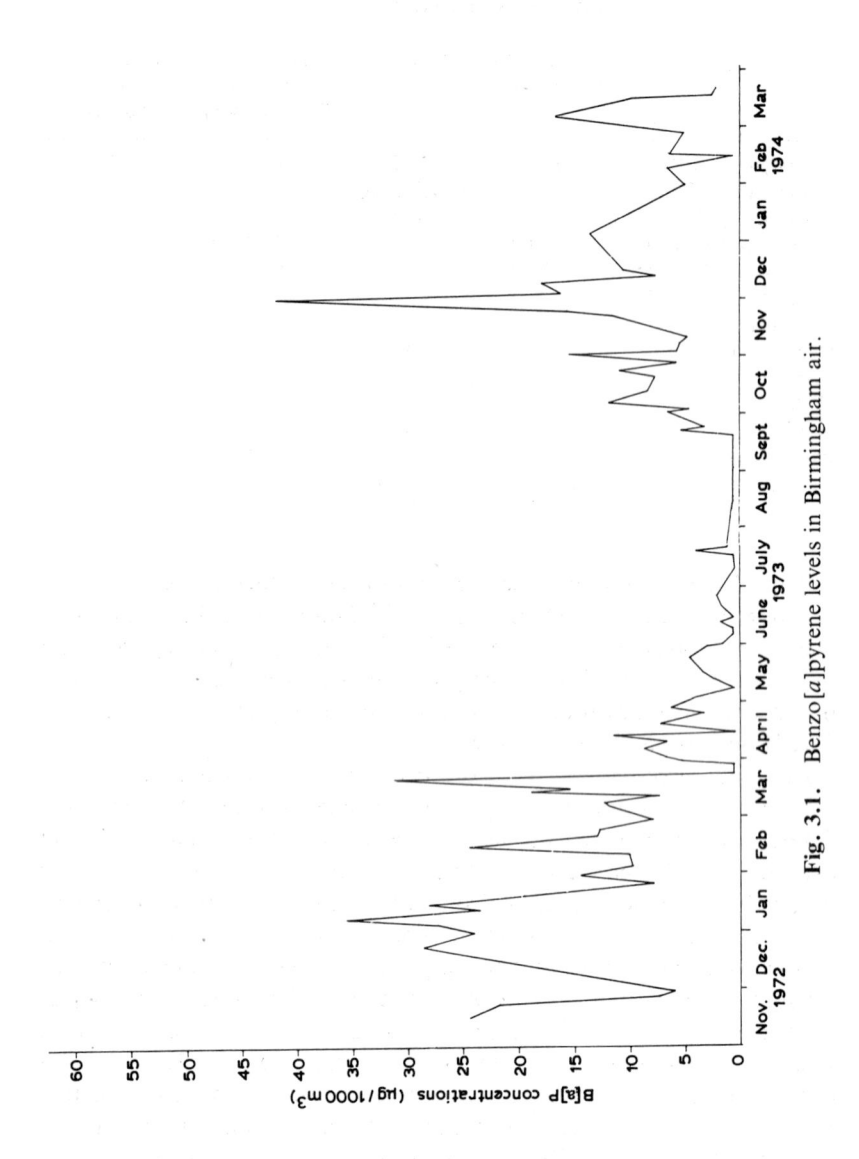

TABLE 3.4
BENZO[a]PYRENE IN FACTORY AIR

Site	Concentration of B[a]P, $ng\,m^{-3}$	
	In factory air	Outside air
Transfermix	19·0	16·6
	16·7	11·5
	ND	9·8
	6·5	2·5
Above extruder	3·2	6·4
	7·9	4·6
	7·3	11·9
	6·0	8·3
Above extruder	8·4	8·3
	5·0	7·6
	12·1	10·9
	8·5	10·9
Tube department	21·3	11·4
Above conveyor	15·7	15·5
	43·0	41·8
	18·8	16·1
Tube department	32·1	41·8
Joining section	ND	16·1
	8·7	7·5
	14·6	10·5
Tyre curing presses	7·9	10·9
	12·8	15·1
	5·3	5·9
	3·7	5·4
Tyre curing presses	9·5	10·9
	9·5	15·1
	3·7	5·9
	4·0	5·4
$63\frac{1}{2}$-in Bagomatic presses	2·2	6·4
	6·1	4·1
	11·1	11·9
	7·0	8·3
Tyre trimming area	23	27·4
	2·2	7·9
	16·2	14·4
	6·4	24·3
	13·4	11·6
	13·5	7·4
	22	18·8
	54	12
	56·8	12
	ND	13
	ND	7·4
	ND	15·4

TABLE 3.5
BENZO[a]PYRENE IN US TYRE FACTORIES

Location	Concentration of B[a]P, ng m^{-3}
Banbury	2·9–32·6
Milling	ND–32·3
Curing	ND–8·8
Inspection	<1–15·3

A further study of PAHs in two rubber factories was made by the BRMA and the Institute of Petroleum in 1979–80.[11] The analytical methods used in this study were more sophisticated than in the earlier study and 14 individual PAHs were measured. Samples were taken at tyre curing, tube curing, tread extrusion, calendering and outside air sites in both factories. None of the carcinogenic PAHs were found to be present in excess, although excesses were found for the two most volatile PAHs (fluoranthene and pyrene) (Table 3.6).

These studies have shown that the conventional carcinogenic PAHs are not sufficiently volatile to be released into the air in significant quantities during rubber processing. Some idea of what might be considered significant can be obtained by comparing the benzo[a]pyrene (B[a]P) concentration found here with amounts reported in other industrial situations, in urban air studies and in cigarette smoke. Cigarette smoke has been shown to contain some 200–12000 ng B[a]P per 100 cigarettes, while the tarry condensate from cigarette smoke contains 1·3 ppm B[a]P.[21] The B[a]P concentration in gasworks retort houses in the UK in the early 1960s was in the range 1400–4800 ng m^{-3} [22] and in coking plants in Czechoslovakia was reported to be in the range 300–35000 ng m^{-3} with an average value 400 m from the ovens of 1800 ng m^{-3}.[23] Clearly the amounts of B[a]P present in the two tyre factories studied above were far less than these figures. For urban air, the National Academy of Sciences report on particulate polycyclic organic matter[24] showed that B[a]P could to some extent be used as a 'marker' for urban air pollution and that each increment of 1 ng m^{-3} B[a]P could be correlated with a 5% increase in lung cancer death rate. Obviously this argument is not directly applicable to the industrial situation, but even with this criterion, no significant excess of B[a]P was detected in the above tests.

However, it has been pointed out that the 14 PAHs measured above comprise only a very small proportion of the total contaminant generated

TABLE 3.6
POLYCYCLIC AROMATIC HYDROCARBONS IN TWO TYRE FACTORIES

PAH	Carcino-genicity	Outside air 1 (January 1980)	Outside air 2 (March 1981)	Concentration of PAH, ng m^{-3} Tyre curing (November 1979–March 1980)	Tube curing (March 1980–April 1980)	Tread extrusion (July 1980)	Calendering (September 1980)
Fluoranthene	−	5	6	9–67	12–139	5–37	2–30
Pyrene	−	3	3	10–119	36–248	12–79	5–66
Benz[a]fluorene	−	2	1	6–16	2–6	2–9	<1–7
Benz[a]anthracene	+	3	11	2–9	1–10	1–20	<1–13
Chrysene	±	5	11	8–25	3–36	5–17	1–27
Benzo[b]fluoranthene	++	4	13	6–8	3–10	1–6	1–4
Benzo[e]pyrene	−	2	6	4–5	2–5	1–4	<1–3
Benzo[a]pyrene	+++	2	7	3–5	2–5	<1–4	<1–2
Dibenz[a,j]anthracene	++	<1	2	<1–1	<1–2	<1–1	<1
Dibenz[a,h]anthracene	+++	<1	1	<1	<1	<1	<1
Indeno[1,2,3-c,d]pyrene	+	1	2	2–3	<1–3	<1–2	<1–1
Benzo[g,h,i]perylene	−	2	5	3–7	2–5	1–4	<1–2
Anthanthrene	−	<1	2	<1–2	<1–1	<1–1	<1–1
Perylene	−	<1	1	<1–2	<1–2	<1–2	<1–1

−, Negative; ±, negative/weakly positive; +, weakly positive; ++, fairly positive; +++, strongly positive.

by hot rubber during curing. In the BRMA/Institute of Petroleum tests, the 14 PAHs formed only about 0·02% of the total fume sample collected. The bulk of the sample (toluene-soluble portion) consisted mainly of phenols, phthalimides and carbazoles with small amounts of sulphur and oxygen heterocycles. These heterocyclic materials could have been present in equivalent amounts to the PAHs.

Very little is known about the biological activity of heterocyclic analogues of PAHs containing O and S atoms or about alkyl substituted PAHs, and further work is required in this area to define the position more clearly.

NITROSAMINES

Nitrosamines have the general structure

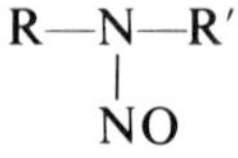

$$R-N-R'$$
$$|$$
$$NO$$

Where the substituents R and R' contain α-hydrogen atoms, the nitrosamine is likely to be carcinogenic:

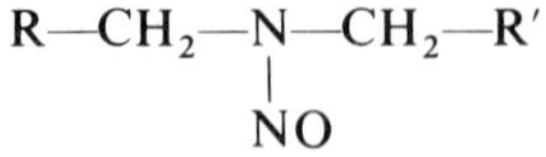

$$R-CH_2-N-CH_2-R'$$
$$|$$
$$NO$$

For example, dimethylnitrosamine is a potent carcinogen in animal tests, producing liver, kidney and lung cancer. The significance of the free α-position can, in certain cases, be demonstrated by replacing one or more of the α-hydrogens by a methyl group; in nitrosopiperidine for example:

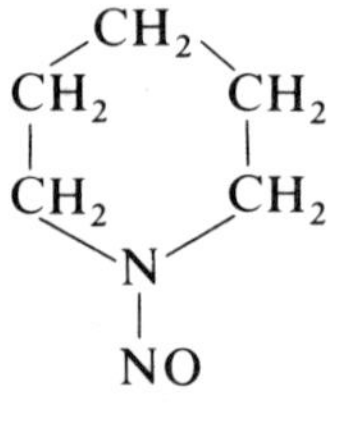

nitrosopiperidine
(strong carcinogen)

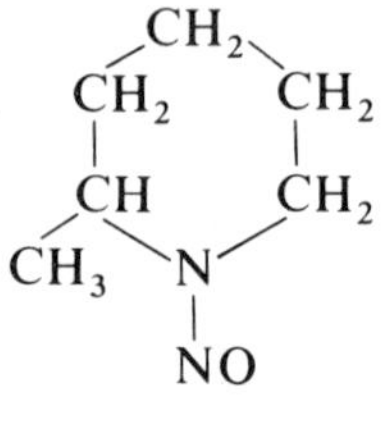

2-methylnitrosopiperidine
(approximately half strength carcinogen)

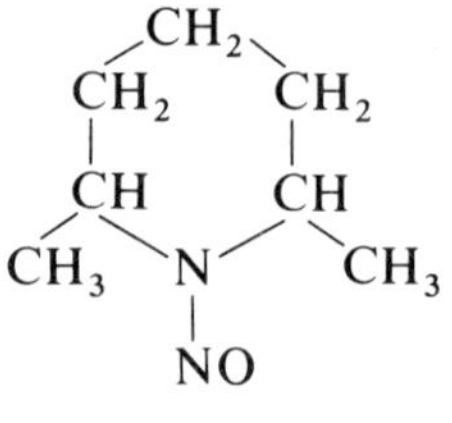

2,6-dimethylnitrosopiperidine
(non-carcinogen)

HAZARDS FROM AROMATIC AMINES, PAHs AND NITROSAMINES 45

In agreement with this mechanism, diphenyl nitrosamine or nitrosodiphenylamine (NDPA) has been shown to be non-carcinogenic in most animal tests,[25,26] although one test on mice gave a positive result.[27,28]

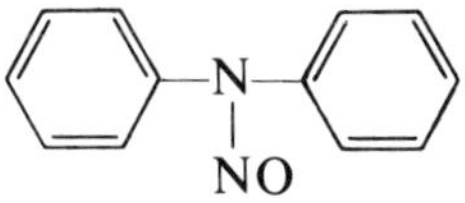

nitrosodiphenylamine

This is of special relevance in the rubber industry where this chemical has been widely used as a retarder (Vulcatard A, Curetard A, Vulkalent A, Retarder J).

Although NDPA is non-carcinogenic, it has now been found that use of this retarder can, under certain circumstances, generate other nitrosamines.[29] This happens when the rubber mix contains materials which generate secondary amines, and the new nitrosamine is formed by transfer of the nitroso group from NDPA to the secondary amine.

Three main groups of rubber chemicals can generate secondary amines:

(*i*) *Thiuramdisulphides:* These produce dialkylamines during curing.

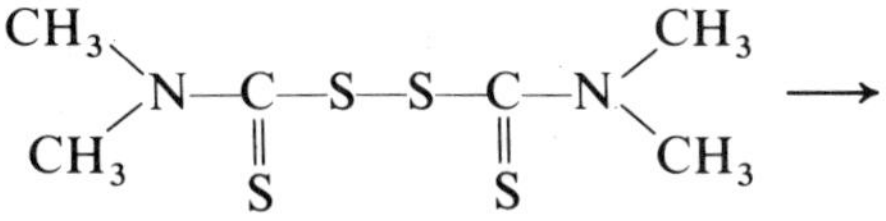

tetramethylthiuramdisulphide

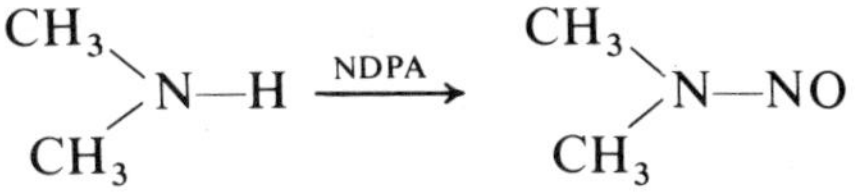

dimethylamine → nitrosodimethylamine

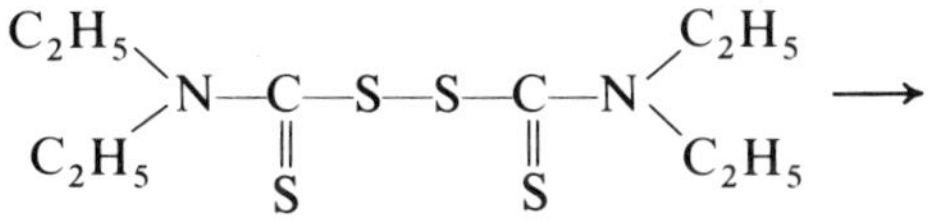

tetraethylthiuramdisulphide

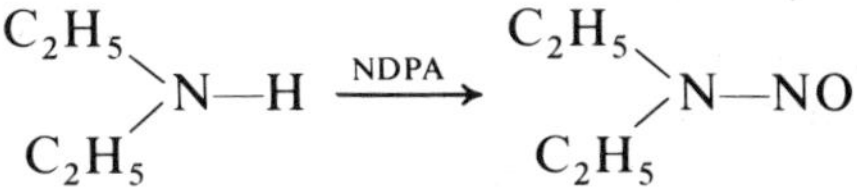

diethylamine → nitrosodiethylamine

(ii) *Dithiocarbamates:*

$$\left(\begin{array}{c}CH_3\\ \\ CH_3\end{array}\!\!\!N\!-\!\!\overset{\overset{\displaystyle S}{\|}}{C}\!-\!S^-\right)_{\!\!2}\!Zn^{++} \longrightarrow \begin{array}{c}CH_3\\ \\ CH_3\end{array}\!\!\!N\!-\!H \xrightarrow{NDPA} \begin{array}{c}CH_3\\ \\ CH_3\end{array}\!\!\!N\!-\!NO$$

zinc diethyldithiocarbamate   dimethylamine   nitrosodimethylamine

(iii) *Sulphenamides:*

[benzothiazole]–C–S–N(morpholine) ⟶ H–N(morpholine) $\xrightarrow{NDPA}$ NO–N(morpholine)

morpholine   *N*-nitrosomorpholine

Nitrosodimethylamine, nitrosodiethylamine, nitrosodibutylamine, *N*-nitrosomorpholine, *N*-nitrosopiperidine and *N*-nitrosopyrrolidine have all been shown to be carcinogenic in animal tests.[30]

Some results obtained by the author from various processing operations are shown in Table 3.7. It can be seen that the highest personal concentrations measured were on injection moulding operations, with up to 380 μg m^{-3} nitrosomorpholine and up to 90 μg m^{-3} dimethylnitrosamine. When NDPA was replaced by an alternative retarder, these personal concentrations fell to around 1 μg m^{-3}. It is possible that the traces of nitrogen oxides in the atmosphere may produce this background concentration of nitrosamines. Compounds with NDPA but no source of secondary amines also produced background concentrations of nitrosamines of the order of 1 μg m^{-3} or less.

The nitrites used in high temperature salt bath cures may also cause nitrosation of any secondary amines which are produced by the rubber compound and relatively high nitrosamine concentrations were found in the direct fumes from a salt bath operation. However, these fumes are generally well controlled, and the concentrations found at working positions were low.

In absolute terms, compared with the atmospheric concentrations of contaminants such as monomers, solvents, etc., all the nitrosamine concentrations found in this work were relatively low. Since little or no information exists on the amounts which would be likely to cause cancer, however, it is difficult to predict whether even these small amounts would cause harmful effects. At 1·0 μg m^{-3} a man working moderately hard

TABLE 3.7
CONCENTRATIONS OF NITROSAMINES IN RUBBER PROCESSING OPERATIONS

Process	Sample	\multicolumn{3}{c}{Nitrosamine concentration, $\mu g\,m^{-3}$}
		NDMA	NDEA	NMor
Injection moulding, compound containing NDPA, TMTD and Sulfasan R, Temperature 175 °C	Personal sample press operator	90	—	380
	Personal sample, press operator	36	—	120
	Static sample, in direct fumes	1060	18	4700
	Static sample, in direct fumes	520	10	2000
As above, after replacing NDPA by alternative retarder	Personal sample, press operator	0·7	—	1·1
	Static sample, in direct fumes	1·9	—	8·7
EPDM compound containing NDPA but no secondary amines	Static sample, in direct fumes	0·5	—	0·2
Brake hose manufacture, compound containing NDPA but no secondary amines	Extruder head	0·6	0·2	1·0
	Mill site	0·2	0·1	0·4
	Curing pan	0·1	—	0·1
Conveyor belt manufacture, compound containing NDPA, TMTD and MBTS	Exit from calender	29	1	8·5
	Exit from curing pans	230	6	37
Nitrite/nitrate salt bath curing of rubber strip containing TMTD, ZDC, but no NDPA	In direct fumes from rubber at exit from salt bath	150	210	720
	Personal sample on extruder operator	1·1	1·1	0·6
	Personal sample on operator coiling strip	1·6	1·7	0·6

would breathe in 10 μg of nitrosamines over an 8-h shift. To put this figure in perspective, the normal intake of dialkylnitrosamines from foodstuffs in the UK is about 1 μg per week, with a further 3 μg per week of heterocyclic nitrosamines.[31] It should be noted, however, that the routes of exposure here are different. Cigarette smoke also contains nitrosamines. The mainstream smoke contains 0·005–0·065 μg nitrosodimethylamine (NDMA) per cigarette, while sidestream smoke contains 0·680–1·770 μg NDMA per cigarette.[32] Other nitrosamines are also present. Smoke-filled rooms such as bars and discotheques contain 0·090–0·240 μg NDMA m^{-3}.[33]

With this background of doubt about the significance of the concentrations of nitrosamines which have been found in the industry, many companies have decided to replace NDPA with other retarders, with the result that it has now largely been removed from rubber processing operations in the UK and other countries.

REFERENCES

1. Crowther, A. F., The investigation into the constitution of the paraldehyde–α-naphthylamine condensation product, *Research Project No. 411*, Cambridge University, 1941; quoted in *Cassidy and Wright versus ICI and Dunlop*, High Court of Justice, 1970, Documents presented, Vol. 3 (contemporary reports), p. 8.
2. Munn, A., Bladder cancer and carcinogenic impurities in rubber additives, *Rubber Industry* (February 1974), pp. 19–20.
3. Strafford, N., *Internal ICI Report SOM 463*, 20 June 1949; quoted in *Cassidy and Wright versus ICI and Dunlop*, High Court of Justice, 1970, Documents presented, Vol. 9 (Miscellaneous), p. 37.
4. Clayson, D. B., Occupational bladder cancer, *Preventative Medicine* (1976), **5**, 228–44.
5. Williams, M. H. C., Occupational tumours of the bladder, in *Cancer*, Vol. 3, ed. R. W. Raven, Butterworths, London, 1958, Chapter 15.
6. *Nonox D and Nonox CI: Investigation of Hazards in Rubber Processing*, ICI Technical Information, PC/R C-60, December 1965.
7. Kummer, R. and Tordoir, W. F., Phenyl β-naphthylamine (PBNA)—another carcinogenic agent? *T. soc. Geneesk.* (1975), **53**, 415.
8. Batten, P. L. and Hathway, D. E., Dephenylation of N-phenyl 2-naphthylamine in dogs and its possible oncogenic implications. *Brit. J. Cancer* (1977), **35**, 342–6.
9. Walpole, A. L., Williams, M. H. C. and Roberts, D. C., The carcinogenic action of 4-aminodiphenyl and 3,3′-dimethyl-4-aminodiphenyl, *Brit. J. Indust. Med.* (1952), **9**, 255.
10. Parkes, H. G., Detection and prevention of health hazards in the rubber industry, *Plastics & Rubber Processing* (December 1977), p. 150.
11. *BRMA Health Bulletin No. 28: Aromatic Oils*, British Rubber Manufacturers' Association, Birmingham, 11 February 1983.
12. *Safety Data and Information on Dutrex Aromatic Processing Oils*, Shell (Chemicals) plc, Ref. SDJ/3/78, 1978.
13. Vickers, H. R., Oil and the skin, *Proc. Inst. Petroleum 1976 Annual Conference: Health and Safety in the Oil Industry*, Heyden & Son, London, 1977, Chapter 9, pp. 123–4.
14. Monson, R. R. and Fine, L. J., Cancer mortality and morbidity among rubber workers, *J. National Cancer Inst.* (1978), **61**, 1047–53.
15. Rivin, D. and Smith, R. G., Environmental health aspects of carbon black, *Rubber Chemistry and Technology* (1982), **55**, 707–61.

16. Buddingh, F., Bailey, M. J., Wells, B. and Haesemeyer, J., Physiological significance of benzo(a)pyrene adsorbed to carbon blacks: elution studies, AHH determinations, *Amer. ind. Hyg. Assoc. J.* (1981), **42**, 503–9.
17. Nau, C. A., Neal, J., Stembridge, V. A. and Cooley, R. N., Physiological effects of carbon black, IV: Inhalation, *Arch. Envir. Health* (1962), **4**, 415–31.
18. Robertson, J. M. and Ingalls, T. H., A mortality study of carbon black workers in the United States from 1935 to 1974, *Arch. Envir. Health* (1980), **35**, 181–6.
19. Nutt, A. R., Measurement of some potentially hazardous materials in the atmosphere of rubber factories, *Envir. Health Perspectives* (1976), **17**, 117–23.
20. Williams, T. M., Harris, R. L., Arp, E. W., Symons, M. J. and Van Ert, M. D., Worker exposure to chemical agents in the manufacture of rubber tyres and tubes: Particulates. *Amer. ind. Hyg. Assoc. J.* (1980), **41**, 204–11.
21. *IARC Monographs on the Evaluation of Carcinogenic Risk of the Chemical to Man*, Vol. 3, International Agency for Research on Cancer, Lyon, France, 1973, p. 96.
22. Lawther, P. J., Commins, B. T. and Weller, R. E., A study of the concentrations of polycyclic aromatic hydrocarbons in gas works retort houses, *Brit. J. Ind. Med.* (1965), **22**, 13.
23. Masek, V., Benzo(a)pyrene in the workplace atmosphere of coal and pitch coking plants, *J. Occupational Med.* (1971), **13**, 193.
24. US National Research Council (Division of Medical Sciences), Committee on Biological Effects of Atmospheric Pollutants, *Particulate Polycyclic Organic Matter*, National Academy of Sciences, Washington, 1972, p. 246.
25. Druckrey, H., et al., *Zeitschr. f. Krebsforschg.* (1967), **69**, 100.
26. Boyland, E., et al., *European J. Cancer* (1968), **4**, 233–9.
27. *US National Technical Information Service, Public Bulletin No. 223*, 1978 p. 159.
28. Cardy, R. H., Lijinsky, W. and Hildebrandt, P. K., Neoplastic and non-neoplastic urinary bladder lesions induced in Fischer 344 rats and B6C3 F$_1$ hybrid mice by N-nitrosodiphenylamine, *Ecotox Environ. Safety* (1979), **3**, 29–35.
29. Fajen, J. M., Carson, G. A. and Rounbehler, D. P., N-Nitrosamines in the rubber and tyre industry, *Science* (1979), **205**, 1262–4.
30. *IARC Monographs on the Evaluation of Carcinogenic Risk of the Chemical to Man*, Vol. 17, International Agency for Research on Cancer, Lyon, France.
31. Gough, T. A., Webb, K. S. and Coleman, R. F., Estimate of the volatile nitrosamine content of UK food, *Nature* (1978), **272**, 161–3.
32. Brunneman, K. D. and Hoffmann, D., *Environmental Aspects of N-nitroso Compounds*, eds E. A. Walker, M. Castegnaro, L. Griciute and R. E. Lyle, International Agency for Research on Cancer, Lyon, France, 1978.
33. Brunneman, K. D., Adams, J. D., Ho, D. P. S. and Hoffman, D., The influence of tobacco smoke on indoor atmospheres. II. Volatile and tobacco specific nitrosamines in main and sidestream smoke and their contribution to indoor pollution, *Cancer Research* (1977), **37**, 3218–22.
34. Hodge, H. C., unpublished work at University of California, San Francisco.

PART II

TOXICITY OF RUBBER CHEMICALS

4

Natural and Synthetic Rubbers

The polymers used in the rubber industry do not themselves give rise to any health problems. The high molecular weight and chemical stability of these polymers are generally sufficient to prevent significant absorption taking place by any route, even when the polymers are used for such medical applications as catheter tubes, implants, etc.

However, the polymers used in the industry sometimes contain trace impurities remaining from the production process and may also have additives such as stabilisers and both these classes of chemical may need consideration for possible toxic hazard.

MONOMERS

Traces of the monomer or monomers used to produce the polymer may remain in commercial polymer supplies. Some of the possible monomers, together with their threshold limit values are listed in Table 4.1. The amount of these monomers present in solid polymer supplies is usually less than 1 ppm by weight, and it is unusual for the TLV to be reached during normal processing. However, the author has measured concentrations of acrylonitrile between 1 and 2 ppm close to the mill rolls during milling of nitrile rubber. Acrylonitrile is now considered to be carcinogenic and checks should be made during work with nitrile rubbers to ensure that exposures to acrylonitrile are well below the TLV of 2 ppm.

Some polymer latices may contain much higher concentrations of monomer, e.g. 5000 ppm by weight of chloroprene monomer in a polychloroprene latex.[1] Again, work with these latices should be checked to ensure that exposure to the monomer does not exceed the TLV.

The specialist polymers may also need checking for residue content. For

TABLE 1
MONOMERS FOUND IN THE RUBBER INDUSTRY

Monomer	TLV–TWA
Acrylonitrile	2 (ACGIH, HSE), 1 (BRMA)
Butadiene	1 000 (ACGIH)
Chloroprene	10 (ACGIH)
Divinylbenzene	10 (ACGIH)
Ethylidene norbornene	6 (C) (ACGIH)
Isoprene	Not listed
Vinyl chloride	5 (HSE)

instance, millable polyurethane rubbers may contain free isocyanate, some silicone polymers may contain species with oestrogenic activity (information supplied by Dow Corning Ltd, 23 May 1980) and chlorosulphonated polyethylene may contain carbon tetrachloride.

POLYMER ADDITIVES

Some polymer supplies contain additives incorporated by the polymer manufacturer. The additives can be grouped into certain classes.

Stabilisers
The identity of any stabiliser present should be ascertained from the polymer supplier. Until a few years ago PBN was a popular stabiliser, but it has now largely been abandoned by polymer manufacturers in the USA and Western Europe. Some polymer supplies from Eastern Europe may still contain PBN, and manufacturers using these polymers should ensure that any traces of free β-naphthylamine will not cause a cancer risk (see Chapter 3).

Oil Extenders and Plasticisers
Some grades of polymer incorporate mineral oil (oil-extended polymer) or plasticiser. For instance, SBR 1712 contains 37·5% aromatic oil, some EPDMs contain paraffinic oils and some nitrile rubbers contain ester plasticisers such as dioctylphthalate. The potential hazards of the aromatic oils have been considered in detail in Chapter 3.

Crosslinking Agents
Some polymers are supplied in a precompounded form, and again the supplier's advice should be sought on the identity and potential health hazards of the additives. For example, some grades of silicone rubber incorporate the highly toxic dibutyl tin dilaurate.

REFERENCE

1. Nutt, A. R., Measurement of some potentially hazardous materials in the atmosphere of rubber factories, *Envir. Health Perspectives* (1976), **17**, 117–23.

5

Reinforcing Agents, Activators and Fillers

CARBON BLACK

Carbon black was originally manufactured by the 'channel' process, but this method is now virtually unused and carbon black is today made largely by the furnace process (more than 90% of production) with a smaller amount of thermal black and a very small quantity of lamp black. The particle size of furnace black is between 25 and 100 nm. Two main health hazards have been suggested for exposure to carbon black. Since the particles are very small and are likely to penetrate into the lower lung spaces, changes in lung function are an obvious possibility. The blacks also contain carcinogenic polycyclic aromatic hydrocarbons and it has been suggested that they may give rise to a cancer hazard.

Lung Function Changes
There is one well-documented case in the medical literature of a pneumoconiosis caused by carbon black.[1] The man affected worked for 21 years in the carbon black store of a rubber factory followed by 11 years in the calender department. Atmospheric concentrations of carbon black were reported to be very heavy. The pneumoconiosis developed by this man may have been complicated by his earlier pulmonary tuberculosis. None of the other workers in the factory, including one who was employed in the carbon black store for 11 years, developed pneumoconiosis. A number of studies from Czechoslovakia and Rumania have also shown that heavy exposure to carbon black has caused cases of pneumoconiosis.[2] Since Crosbie and coworkers[22] have shown in a study of 506 workers in carbon black production plants in the UK and the USA that there was no evidence of harmful effects on respiratory function it seems likely that prolonged exposure to very heavy concentrations of carbon black, well above the

current TLV ($3.5\,mg\,m^{-3}$) are required to produce this type of pneumoconiotic change.

Two other studies of lung function in groups of carbon black workers have been reported. In the first study[3] a group of 35 carbon black workers in Zagreb, Yugoslavia, were examined over a 6 year period. The mean concentrations of carbon black in the work environment were $8.4\,mg\,m^{-3}$ (total) and $7.5\,mg\,m^{-3}$ (respirable). Both FVC and FEV were found to have declined more rapidly than predicted by comparison with the general population. The drops in FVC and in FEV were 4 and 3 times the predicted drops, respectively. 17% of the subjects were found to have experienced radiological lung changes, described as interstitial fibrosis of a discrete reticular and finely nodular structure mainly in the middle and basal pulmonary segments.

The second study involved 65 men exposed to dust in the Banbury area of three tyre plants in the USA and 189 control workers.[4] Carbon black constituted 70% of the non-rubber materials added to the Banbury. The mean dust levels at each plant were below the current TLV for carbon black of $3.5\,mg\,m^{-3}$. There was no significant difference in FEV or FVC between the control and the processing workers group, but there was a significant decrease in the FEV/FVC ratio for those workers with more than 10 years' exposure.

Cancer

The presence of carcinogenic polycyclic aromatic hydrocarbons in carbon black has been demonstrated since the 1950s.[5,6] More PAHs are present in furnace black than in channel black. Some detailed recent analyses were given in Chapter 3.

It appears that these PAHs are strongly adsorbed on the carbon black surface and are not released in significant quantities by body fluids. Nau carried out a series of studies which showed that carbon black itself, carbon black which had been extracted with benzene and extracted carbon black which had had a known carcinogen (methyl cholanthrene) added to it did not cause cancer when fed to mice.[7] On the other hand, both the benzene extract of carbon black and the known carcinogen caused stomach cancer when mixed with the feed. Similarly the first series of materials, when painted onto the skin of mice, rabbits and monkeys did not cause skin cancer, whereas the benzene extract and the carcinogen did.[8] Subcutaneous and intraperitoneal routes of absorption were also investigated with similar results.[9] Inhalation of furnace carbon black dust concentrations of up to $56.5\,mg\,m^{-3}$ for prolonged periods (sometimes the lifetime of the test animal), using hamsters, mice, guinea pigs and monkeys gave no evidence of cancer.[10]

Two epidemiological studies have been carried out on carbon black production workers. Ingalls[11] found that the cancer mortality and morbidity rates in a population of 476 carbon black workers and 322 other workers employed on US production plants during the 10 year period 1939–1949 was no higher than would be expected in similar industrial populations. The study was subsequently extended for 7 years, and the cancer rates were actually less than expectation.[12] However the numbers in the study were small and the duration of exposure was not sufficient to allow firm conclusions to be drawn. Robertson and Ingalls carried out a retrospective cohort study of workers of four US carbon black manufacturers.[13] No exposure data were given for this population. Of the 190 deaths recorded, 29 were due to malignant disease, this number being no greater than expectations based on rates for white males in the states in which they were employed. The 89 deaths from heart disease were again no more than expected. The evidence to date therefore suggests that carbon black, at the exposure concentrations which have been present during carbon black production over the past 20 or so years, does not cause lung or other types of cancer.

Finally, however, in addition to these chronic potential health hazards, carbon black may cause an acute hazard when stored in confined areas, due to release of adsorbed carbon monoxide.

MINERAL FILLERS, ACTIVATORS AND DUSTING AGENTS

Most of the materials in this category are relatively free from health problems. Many of them have, however, been used in forms which cause atmospheric dust, often in unacceptable concentrations. Even the least hazardous of these materials must be controlled to below the nuisance dust TLV of $10 \, \text{mg m}^{-3}$. The materials in Table 5.1 can be regarded as nuisance dusts with minimal potential health hazards.

Other members of this class have been shown to produce physiological effects, and require more careful handling.

Talc

Talc is a hydrated magnesium silicate with a plate-like structure. It is widely used as a dusting powder which prevents self-adhesion between tacky rubber surfaces, and acts as a lubricant in extrusion and other processes. The exact composition of the natural mineral varies with the source, and

TABLE 5.1
NUISANCE DUSTS

Aluminium hydroxide
Aluminium silicate
Barium sulphate
Calcium carbonate (Whiting)
Calcium silicate
Clays with less than 1% free crystalline silica (i.e. most china clays)
Lithopone
Magnesium oxide
Titanium oxide
Zinc oxide
Zinc stearate

some supplies of talc, especially from North American mines, contain asbestos-like fibres (tremolite, anthophyllite, etc.). Wherever possible, fibre-free supplies should be preferred. It has been shown that excessive inhalation of talc dust over a period of years can cause respiratory disease. This may take the form of a persistent cough, sputum and breathlessness as a result of airway irritation and obstruction. A study of a group of rubber workers exposed to non-fibrous talc in two US rubber plants[14] showed that they had a greater tendency towards mucous hypersecretion, chronic bronchitis and wheezing, and had a greater loss of $FEV_{1.0}$ than a control group not exposed to talc, carbon black or curing fumes. The respirable dust concentrations in the areas where these workers were employed were $0.60\,mg\,m^{-3}$ (inner tube splicing), $1.41\,mg\,m^{-3}$ (inner tube curing) and $3.55\,mg\,m^{-3}$ (rubber band area). The authors concluded that exposure at the present respirable TLV of $1.0\,mg\,m^{-3}$ causes lung impairment and increases respiratory morbidity. They recommend a reduction in the respirable TLV to at least $0.5\,mg\,m^{-3}$ and preferably $0.25\,mg\,m^{-3}$.

At higher concentrations where substantial quantities of dust enter the lungs more serious lung fibrosis can occur.[15-17] The term 'talcosis' has been used to describe this form of pneumoconiosis. A comprehensive review of the biological effects of talc has been published.[18]

Mica
Mica is less used by the rubber industry than talc, and it has not been subjected to such extensive study. However, it is likely that it has similar effects to talc.

Minerals With More Than 1% of Crystalline Silica
Exposure to respirable sizes of crystalline silica (quartz) causes silicosis—a progressive form of lung fibrosis. Some of the naturally occurring fillers

used in the industry contain significant quantities of crystalline silica. In this context, quantities greater than 1% can be regarded as significant. Perhaps the most common of these fillers are the 'ball-clays' which may contain 10–20% free crystalline silica. Kieselguhr (diatomaceous earth) may occasionally contain crystalline silica, especially when it has been calcined. Although most china clays contain less than 1% free crystalline silica, some supplies may contain more.

Where such materials are used, they should be controlled to below the TLV calculated by using the formulae

$$\text{TLV (total dust)} = \frac{30}{\% \text{ quartz} + 3}, \text{mg m}^{-3}$$

$$\text{TLV (respirable dust)} = \frac{10}{\% \text{ quartz} + 2}, \text{mg m}^{-3}$$

The % quartz figures should be obtained from the respective airborne samples.

Synthetic Amorphous Silicas

Amorphous silica is manufactured by two processes. Hydrated amorphous silica is obtained by precipitation from sodium silicate. Anhydrous silica is made by either hydrolytic decomposition of silicon tetrachloride with air and hydrogen at 100 °C or by dehydrating sodium silicate with alcohol. Although the particle size of these forms of amorphous silica is very small (0·01–0·03 µm), studies of exposed workers have given negative results for silicosis. One study in a German plant manufacturing anhydrous silica from silicon tetrachloride observed 215 workers over a 14 year period.[18] Dust concentrations in the inhaled air varied from 2 to 7 mg m^{-3}. Nine of these workers had been employed on the silica plant for over 10 years. No evidence of silicosis was found. A second study observed 76 men employed in a plant manufacturing hydrated silica during the period 1941–1959.[19] Dust concentrations varied from 0·4 to 205 mg m^{-3}. No evidence of silicosis or other pulmonary disease was found.

Resorcinol

This chemical is commonly used in bonding systems for increasing the adhesion of rubber to textile components. It has the structure

It is usually supplied in flake form and has an LD_{50} (oral, rat) of 300 mg kg^{-1}. Resorcinol is an irritant to skin, eyes and respiratory tract and can cause skin sensitisation. The TLV for resorcinol (ACGIH, 1983) is 10 ppm (45 mg m^{-3}). However, use in the rubber industry has produced complaints of eye and throat irritation at concentrations between 1 and 10 mg m^{-3}, and the BRMA has recommended a working limit of 1 mg m^{-3}.[21]

Resorcinol is a systematic poison, acting on the central nervous system. Since it is absorbed through the skin, contact with solutions containing it must be avoided.

REFERENCES

1. Miller, A. A. and Ranswen, F., Carbon pneumoconiosis, *Brit. J. Ind. Med.* (1961), **18**, 103–13.
2. Concarla, A., Cornea, G., Dengel, H., Gabor, S., Milea, M. and Papilian, V., *Int. Arch. Occup. Environ. Health* (1976), **36**, 217.
3. Valik, F., Beritic-Stahuljak, D. and Mak, B., *Int. Arch. Arbeitsmedezinishe* (1975), **34**, 57–63.
4. Fine, L. J. and Peters, J. M., Studies of respiratory morbidity in processing workers, *Arch. Envir. Health* (1976), **31**, 136–40.
5. Falk, H. L., Steiner, P. E., Goldfein, S., Brescow, A. and Hykes, R., Carcinogenic hydrocarbons and related compounds in processed rubber, *Cancer Research* (1951), **11**, 318–24.
6. Falk, H. L. and Steiner, P. E., The identification of aromatic polycyclic hydrocarbon in carbon black, *Cancer Research* (1952), **12**, 30–9.
7. Nau, C. A., Neal, J. and Stembridge, V., A study of the physiological effects of carbon black, I: Ingestion, *AMA Arch. Ind. Health* (1958), **17**, 21–8.
8. Nau, C. A., Neal, J. and Stembridge, V., A study of the physiological properties of carbon black, II: Skin contact, *AMA Arch. Ind. Health* (1960), **18**, 511–20.
9. Nau, C. A., Neal, J. and Stembridge, V., A study of the physiological effects of carbon black, III: Absorption and elution potential, subcutaneous injections, *Arch. Envir. Health* (1960), **1**, 512–33.
10. Nau, C. A., Neal, J., Stembridge, V. and Cooley, R. N., Physiological effects of carbon black, IV: Inhalation, *Arch. Envir. Health* (1962), **4**, 415–31.
11. Ingalls, T. H., Incidence of cancer in the carbon black industry, *Arch. Ind. Hyg.* (1950), **1**, 662–76.
12. Ingalls, T. H. and Risquez-Irribaren, R., Periodic research for cancer in the carbon black industry, *Arch. Envir. Health* (1961), **2**, 429–33.
13. Robertson, J. M. and Ingalls, T. H., A mortality study of carbon black workers in the United States from 1935 to 1974, *Arch. Envir. Health* (1980), **35**, 181–6.

14. Fine, L. J., Peters, J. M., Burgess, W. A. and Berardnis, L. J. D., Studies of respiratory morbidity in rubber workers, IV: Respiratory morbidity in talc workers, *Arch. Envir. Health* (1976), **31**,195–200.
15. *Control of Exposure to Talc Dust*, Guidance Note, Environmental Health: EH 32, Health and Safety Executive, London, July 1982.
16. Leophante, P., Fabres, J., Pous, J., Albarede, J. L. and Delaude, A., Les pneumoconioses par le talc, *Archives des Maladies Professionelles, de Médécine du Travail et de Sécurité Sociale* (1976), **37**, 513–32.
17. Kleinfeld, M., Messite, J., Shapiro, L., Sweneicki, R. and Sarfaty, J., Lung function changes in talc pneumoconiosis, *J. Occupational Med.* (1965), **7**, 12–17.
18. Hildick-Smith, G. Y., The biology of talc, *Brit. J. Ind. Med.* (1976), **33**, 217–29.
19. Volk, H., The health of workers in a plant making highly dispersed silica, *Arch. Envir. Health* (1960), **1**, 125–8.
20. Plunkett, E. R. and Dewitt, B. J., Occupational exposure to Hi-Sil and Silene, *Arch. Envir. Health* (1962), **5**, 75–8.
21. *Toxicity and Safe Handling of Rubber Chemicals*, British Rubber Manufacturers' Association Code of Practice, BRMA, Birmingham, 1978.
22. Crosbie, W. A., Cox, R. A. F., Leblanc, J. V., Cooper, D. and Thomas, W. C., A respiratory survey on carbon black workers in the UK and USA, 1980 (unpublished).

6

Curing Agents

SULPHUR

Sulphur is a material of low toxicity. The dust is slightly irritating to skin and eyes. One study in the United States found no adverse effects due to inhalation of the dust,[1] though an Italian report found cases of 'thiopneumoconiosis' and bronchitis with emphysema.[2,3] It is likely that exposure levels in the rubber industry are well below those which would cause this type of problem.

DITHIODIMORPHOLINE

$$O\diagup\!\!\diagdown N-S-S-N\diagup\!\!\diagdown O$$

Proprietary names:	Sulfasan R—Monsanto
Physical form:	Off-white powder (also available dust suppressed)
Acute toxicity:	LD_{50} (oral, rat) 5600 mg kg^{-1}
	LD_{50} (dermal, rabbit) greater than 5010 mg kg^{-1}
	On heating releases morpholine
Skin and eye irritation:	Potent primary skin irritant (man)
	Skin sensitiser (man)
	Eye irritant (man)

DICUMYL PEROXIDE

$$\text{Ph}-\underset{\underset{CH_3}{|}}{\overset{\overset{CH_3}{|}}{C}}-O-O-\underset{\underset{CH_3}{|}}{\overset{\overset{CH_3}{|}}{C}}-\text{Ph}$$

Proprietary names:	95% active material—Dicup R, Dicup T (Hercules); Perkadox SB, Perkadox BC (AKZO); Luperco 500R, Luperco 500T (Pennwalt)
	40% active material—Dicup 40C, Dicup KE (Hercules); Perkadox BC-40B, Perkadox BC-40K (AKZO); Luperco 540-C, Luperco 540-KE (Pennwalt)
Physical form:	White powder
Acute toxicity:	LD_{50} (oral, rat) 4100 mg kg^{-1}
Skin and eye irritation:	Skin and eye irritant (man)

Other Comments
Produces acetophenone and α,α-dimethyl benzyl alcohol at cure temperatures.[4]

DITERT. BUTYLPEROXIDE

$$CH_3-\underset{\underset{CH_3}{|}}{\overset{\overset{CH_3}{|}}{C}}-O-O-\underset{\underset{CH_3}{|}}{\overset{\overset{CH_3}{|}}{C}}-CH_3$$

Proprietary name:	Triganox B—AKZO
Physical form:	Colourless liquid
Acute toxicity:	LD_{50} (oral, rat) greater than 25 000 mg kg^{-1}
Skin and eye irritation:	Mildly irritating to eyes (rabbit)
	Mildly irritating to skin (rabbit, man)

Other Comments
Produces acetone and tert.-butanol at cure temperatures.[4]

TERT. BUTYLPEROXYBENZOATE

$$CH_3-\underset{\underset{CH_3}{|}}{\overset{\overset{CH_3}{|}}{C}}-O-O-\underset{\underset{O}{\|}}{C}-C_6H_5$$

Proprietary names:	95% active material—Triganox C (AKZO); Luperox P (Pennwalt)
	50% active material—Triganox C50D (AKZO)
Physical form:	95% active material—colourless liquid
	50% active material—white powder
Acute toxicity:	LD_{50} (oral, rat) 3639 mg kg^{-1}
Skin and eye irritation:	Mildly irritating to eyes (rabbit)
	Mildly irritating to skin (rabbit)

BIS(TERT. BUTYLPEROXYISOPROPYL) BENZENE

$$CH_3-\underset{\underset{CH_3}{|}}{\overset{\overset{CH_3}{|}}{C}}-O-O-\underset{\underset{CH_3}{|}}{\overset{\overset{CH_3}{|}}{C}}-C_6H_4-\underset{\underset{CH_3}{|}}{\overset{\overset{CH_3}{|}}{C}}-O-O-\underset{\underset{CH_3}{|}}{\overset{\overset{CH_3}{|}}{C}}-CH_3$$

Proprietary names:	95% active material—Perkadox 14-90 (AKZO)
	40% active material—Perkadox 14-40K (AKZO)
Physical form:	White powder
Acute toxicity:	LD_{50} (oral, rat) 2300 mg kg^{-1}
Skin and eye irritation:	Slightly irritating to eyes (rabbit)
	Mildly irritating to skin (rabbit)

N,N'DICINNAMYLIDENE-1,6-HEXANEDIAMINE

$$\text{Ph}-CH=CH-CH=N-(CH_2)_6-N=CH-CH=CH-\text{Ph}$$

Proprietary name: Diak No. 3—DuPont
Physical form: Coarse, tan powder
Acute toxicity: Lethal concentration (rats) 6·2 mg litre^{-1} of air
Skin and eye irritation: Skin and eye irritant (man)
Dust irritating when inhaled

HEXAMETHYLENEDIAMINE CARBAMATE

$$H_3N^+-(CH_2)_6-N{<}^{CO_2^-}_{H}$$

Proprietary name: Diak No. 1—DuPont
Physical form: Very fine, white powder
Acute toxicity: LD$_{50}$ (oral, rat) 2875 mg kg^{-1}
Skin and eye irritation: Skin and eye irritant (man)

1,4-DIBENZOYL-p-BENZOQUINONE DIOXIME

$$\text{Ph}-\overset{O}{\underset{}{C}}-O-N=\text{C}_6\text{H}_4=N-O-\overset{O}{\underset{}{C}}-\text{Ph}$$

Proprietary name: Dibenzo GMF—Uniroyal
Physical form: Brown powder
Acute toxicity: LD$_{50}$ (oral, rat) greater than 10 000 mg kg^{-1}
Skin and eye irritation: Possible skin and eye irritant

REFERENCES

1. Pinta, S. S., Brown, R. A. and Carlton, B. H., Study of industrial workers exposed to sulphur dust, *J. Ind. Hyg. & Toxicology* (1943), **25**, 149–51.
2. Schiavina, G. P., Thiopneumoconiosis, *Rassegria di Medicine Industriale* (1941), **12**, 173–85.
3. Frada, G., Mentesana, G. and Azzaro, V., Observations and considerations on the pathology of workers in a sulphur mine, *Folia Medica (Naples)* (1957), **40**, 525–43.
4. Werverka, D., Hummel, K. and Inselsbacher, W., *Rubber. Chem. Technol.* (1976), **49**, 1142.

7

Oils, Waxes, Resins and Plasticisers

PROCESSING AND EXTENDER OILS

These oils are divided, somewhat arbitrarily, into three classes, as shown in Table 7.1.

TABLE 7.1

Type	Composition, carbon atoms %		
	Aromatic	*Naphthenic*	*Paraffinic*
Aromatic	40–60	20–40	30–40
Naphthenic	20–25	35–40	40–45
Paraffinic	4	25–30	67–73

The aromatic oils, because of their relatively high content of polycyclic aromatic hydrocarbons, are the most hazardous of these three classes. A detailed account of the topic is given in Chapter 3.

PETROLEUM WAXES

These materials have a low degree of toxicity.

THE CHLORINATED PETROLEUM WAXES

Proprietary name:	Cereclors—ICI
Physical form:	Finely divided powders, some liquids
Acute toxicity:	LD_{50} (oral) greater than 4000 mg kg^{-1}
Skin and eye irritation:	Slightly irritating to eyes (rabbit)
	Not irritating to skin (rats, human industrial experience)
Chronic toxicity:	90-day feeding studies have been carried out on one of the liquid grades:
	No-effect levels were 900 ppm (dogs), 500 ppm (rats)
	At greater dosage both species showed enlargement of the liver and an increase in serum glutamic pyruvate transaminase
	At very high dosage (5000 ppm) some heart damage was found
	Feeding studies with ^{14}C-labelled materials showed that no tissue accumulation occurred

TACKIFYING RESINS

Aliphatic Petroleum Hydrocarbon Resins
These materials have a low degree of hazard.

Aromatic Petroleum Hydrocarbon Resins
These materials may contain significant concentrations of polycyclic aromatic hydrocarbons. The B[a]P concentration in one resin was measured as 10 ppm.[1] Care should be taken in handling to avoid skin contact or inhalation of dust.

Coumarone–Indene Resins
These materials are obtained by coal tar distillation and consist of polyindenes. They may contain some PAHs (e.g. 1–2 ppm benzo[a]pyrene), but are less likely to be hazardous than the aromatic petroleum resins. They should nevertheless be handled carefully.

ESTER PLASTICISERS

Phthalate Esters
Commonly used phthalate esters are:

(i) dibutyl phthalate;
(ii) diisobutyl phthalate;
(iii) di-2-ethylhexyl phthalate (dioctyl phthalate);
(iv) dialphanyl phthalate.

These materials generally have a low toxicity and are non-irritant to skin and eyes. Oral LD_{50} values in the rabbit for the commonly used materials are all greater than 8000 mg kg^{-1}. Skin LD_{50} values in the rabbit are all greater than 10 000 mg kg^{-1}. Dioctyl phthalate has produced liver cancers when fed to mice and rats at very high dose levels (up to 1·2% for 12 months)[2] but the exposure levels concerned were thought to be so high as to make these results irrelevant to human experience. Previous studies at lower feeding levels (0·4–0·5% in the diet for 2 years) had given negative results for tumours.[3] This plasticiser is still allowed for food contact by the US Food and Drug Administration.

Phosphate Esters
Triaryl Phosphates

$$\left[(CH_3)_n - \underset{}{\bigcirc} - O \right]_3 P=O \qquad n = 0, 1 \text{ or } 2$$

Triphenyl, tricresyl (tritolyl) and trixylyl phosphates have all been used as plasticisers in the rubber industry. They have low acute toxicity (LD_{50} values 5000–30 000 mg kg^{-1}) as measured by conventional LD_{50} animal tests, but some of the isomers cause severe central nervous system damage.

Tricresyl phosphates were originally made from a mixture of cresol isomers, and some serious and large scale outbreaks of poisoning took place some 30 years ago when these phosphates were accidentally incorporated into foodstuffs and drink (e.g. the 'Jamaica Ginger' episode[6]). Tests showed that the active material was the *ortho* isomer, tri-*o*-cresyl phosphate, which caused damage to the central nervous system resulting in muscular weakness and paralysis. Recovery was slow, and some permanent paralysis occurred in the worst affected cases. The minimum paralytic dose in man may be between 10 and 30 mg kg^{-1}.[4]

Tricresyl phosphates derived from *m*- and *p*-cresols are relatively inert, and commercial tricresylphosphate is now made from cresol mixtures containing less than 0·1% *ortho* isomer. Tricresyl phosphate containing less than 0·1% tri-*o*-cresyl phosphate can be safely used with normal handling precautions in the rubber industry, though care should be taken to prevent direct skin contact with the liquid.

Some supplies of trixylyl phosphate may also contain traces of tri-*o*-cresyl phosphates, and users should check with their suppliers that these plasticisers are safe with respect to neurotoxicity.

Alkyl Diaryl Phosphates
Materials such as 2-ethylhexyl-diphenyl-phosphate have found limited use as plasticisers in the rubber industry.

These materials are relatively inert compared with tri-*o*-cresyl-phosphate and should not cause toxicity problems. Direct skin contact with the liquid should nevertheless be avoided.

Adipate and Sebacate Esters
Di-2-ethylhexyl-adipate and di-2-ethylhexyl-sebacate are in use as plasticisers. These materials have a low degree of acute and chronic toxicity. Dioctyl adipate has produced liver cancers when fed to mice and rats at very high dose levels,[5] but the exposure levels concerned were thought to be so high as to make these results irrelevant to human experience (see also di-2-ethylhexyl-phthalate). This plasticiser is still allowed for food contact by the US Food and Drug Administration.

REFERENCES

1. *Information on Escorez 3000 Resin Series*, Esso (Chemicals) Ltd, November 1972.
2. *NTP Technical Report Series 217 on the Carcinogenesis Bioassay of Di-(2-ethylhexyl)phthalate in F344 Rats and B6C3 F_1 Mice*, National Institute of Health Publication No. 82-1773, USA.
3. Carpenter, C. P., Weil, H. F. and Smyth, H. F., *Arch. Ind. Hyg. Occup.* (1953), **8**, 219.
4. Clayton, G. D. and Clayton, F. E. (Eds), *Patty's Industrial Hygiene and Toxicology*, Vol. 2A, 3rd Edn, Wiley–Interscience, New York, 1981, p. 2369.
5. *NTP Technical Report Series 212 on the Carcinogenesis Bioassay of Di-(2-ethylhexyl) Adipate in F344 Rats and B6C3 F_1 Mice*, National Institute of Health Publication No. 81-1768, USA.
6. Patty, F. A. (Ed.), *Industrial Hygiene and Toxicology*, Vol. II, 2nd Edn, Interscience, New York, 1962, p. 1917.

8

Accelerators

Condensation Products

BUTYRALDEHYDE–ANILINE

This material has a complex structure which cannot be represented by a simple chemical formula.

Proprietary name:	Beutene—Uniroyal
Physical form:	Amber to red-brown liquid
Acute toxicity:	LD_{50} (oral, rat) $4180\,mg\,kg^{-1}$
	LD_{50} (dermal, rabbit) greater than $7940\,mg\,kg^{-1}$
	On heating gives fumes which are irritant when inhaled
Skin and eye irritation:	Neither skin nor eye irritant (rabbit)

HEPTALDEHYDE–ANILINE

This material has a complex structure which cannot be represented by a simple chemical formula.

Proprietary name:	Heptene base—Uniroyal
Physical form:	Free flowing liquid
Acute toxicity:	LD_{50} (oral, rat) not available
Skin and eye irritation:	Slightly irritating to skin

FORMALDEHYDE–AMMONIA–ETHYL CHLORIDE

This material has a complex structure which cannot be represented by a simple chemical formula.

Proprietary names:	Trimene base—Uniroyal
	Vulcastad EFA—Vulnax
Physical form:	Dark brown, viscous liquid
Acute toxicity:	LD_{50} (oral, rat) greater than 2000 mg kg^{-1}
Skin and eye irritation:	Severely irritating to skin and eyes

Amines

HEXAMETHYLENE TETRAMINE (HMT)

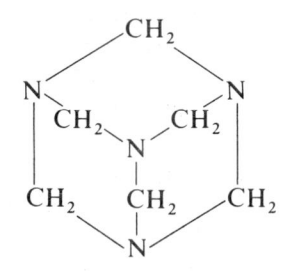

Proprietary names:	Vulkacit H30—Bayer
	HMT, Hexa — various suppliers
Physical form:	White powder
Acute toxicity:	LD_{50} (oral, rat) greater than 10 000 mg kg^{-1}
Skin and eye irritation:	Skin and eye irritant
	Has gained a bad reputation as a skin sensitiser in the synthetic resin industry,[1] where it is used as a curing agent, but has given less trouble in the rubber industry
Chronic toxicity:	Has been used as a medicament in doses up to 5 g per day

Chronic toxicity: No evidence of carcinogenicity in oral feeding tests on mice for 60 weeks and rats for 2 years.[2] Doses in rats were 1% of the drinking water

Dithiocarbamates

ZINC DIMETHYLDITHIOCARBAMATE (ZDMC)

$$\left[\begin{array}{c} CH_3 \\ \diagdown \\ \diagup \\ CH_3 \end{array} N - \overset{\overset{\displaystyle S}{\|}}{C} - S \right]_2 Zn$$

Proprietary names:	Anchor ZDMC—Anchor
	Methasan—Monsanto
	Methazate—Uniroyal
	Robac ZMD—Robinson
	Vulcafor ZDMC—Vulnax
	Vulkacit L—Bayer
Physical form:	White powder
Acute toxicity:	LD_{50} (oral, rat) 500–1400 mg kg^{-1}
	LD_{50} (dermal, rabbit) 5010–7940 mg kg^{-1}
Skin and eye irritation:	Moderately irritating to eyes (rabbit)
	Not irritating to skin (rabbit)
	Not a primary skin irritant (human patch tests)
Chronic toxicity:	Carcinogenicity tests have given contradictory results:
	Rats fed 0·05 of the LD_{50} twice weekly for 22 months gave a positive incidence of tumours[3]
	Rats fed 0·0025–0·025% ZDMC in their diet for 2 years showed no rise in tumour incidence[4]
	Mice fed daily doses of 4·6 mg kg^{-1} in their diet showed no rise in tumour incidence[5]
	On the basis of these results ZDMC is unlikely to be significantly carcinogenic

ZINC DIETHYLDITHIOCARBAMATE (ZDEC)

$$\left[\begin{array}{c} C_2H_5 \\ C_2H_5 \end{array} \!\!\!> N-\overset{\overset{\displaystyle S}{\|}}{C}-S \right]_2 Zn$$

Proprietary names:	Anchor ZDEC—Anchor
	Ekagom 4R—Ugine Kuhlmann
	Ethasan—Monsanto
	Ethazate—Uniroyal
	Ethyl Zimate—Vanderbilt
	Robac ZDC—Robinson
	Vulcafor ZDEC—Vulnax
	Vulkacit LDA—Bayer
Physical form:	White powder
Acute toxicity:	LD_{50} (oral, rat) 3530 mg kg^{-1}
	LD_{50} (dermal, rabbit) 3160 mg kg^{-1}
Skin and eye irritation:	Neither skin nor eye irritant (rabbit)
	Some clinical evidence of sensitisation from contact with finished products, but has caused little problem in industrial use

ZINC DIBUTYLDITHIOCARBAMATE (ZDBC)

$$\left[\begin{array}{c} C_4H_9 \\ C_4H_9 \end{array} \!\!\!> N-\overset{\overset{\displaystyle }{\|}}{\underset{\underset{\displaystyle S}{}}{C}}-S \right]_2 Zn$$

Proprietary names:	Anchor ZDBC—Anchor
	Butazate—Uniroyal
	Robac ZBUD—Robinson
	Vulcafor ZDBC—Vulnax
Physical form:	White powder
Acute toxicity:	LD_{50} (oral, rat) greater than 2000 mg kg^{-1}
Skin and eye irritation:	Neither skin nor eye irritant
Chronic toxicity:	Has given negative results in one carcinogenicity test in mice[5]

ZINC PENTAMETHYLENE DITHIOCARBAMATE (ZPD)

$$\left[CH_2 \diagup \begin{matrix} CH_2-CH_2 \\ CH_2-CH_2 \end{matrix} \diagdown N-\overset{\overset{\displaystyle S}{\|}}{C}-S \right]_2 Zn$$

Proprietary names:	Vulkacit ZP—Bayer
	Robac ZPD—Robinson
Physical form:	Off-white powder
Acute toxicity:	LD_{50} (oral, rat) 3720 mg kg^{-1}
	The dust is irritating to the nose and throat
Skin and eye irritation:	Not irritating to skin (rabbits)
	Severely irritating to eyes

ZINC DIBENZYLDITHIOCARBAMATE (ZBED)

$$\left[\begin{matrix} C_6H_5CH_2 \\ C_6H_5CH_2 \end{matrix} \diagdown N-\overset{\overset{\displaystyle }{}}{\underset{\underset{\displaystyle S}{\|}}{C}}-S \right]_2 Zn$$

Proprietary name:	Robac ZBED—Robinson
Physical form:	Creamy white powder
Acute toxicity:	LD_{50} (oral, rat) greater than 2000 mg kg^{-1}
Skin and eye irritation:	Neither skin nor eye irritant

ZINC ETHYL-PHENYL-DITHIOCARBAMATE

$$\left[\begin{matrix} C_2H_5 \\ C_6H_5 \end{matrix} \diagdown N-\overset{\overset{\displaystyle S}{\|}}{C}-S \right]_2 Zn$$

Proprietary names:	Vulkacit P extra N—Bayer
	Ancazate EPH—Anchor
Physical form:	Yellow–white powder
Acute toxicity	LD_{50} (oral, rat) greater than 10 000 mg kg^{-1}
	The dust is irritating to the nose and throat
Skin and eye irritation:	Slightly irritating to eyes (rabbit)
	Not irritating to skin (rabbit)

NICKEL DIBUTYLDITHIOCARBAMATE (NBC)

$$\left[\begin{array}{c} C_4H_9 \\ C_4H_9 \end{array} \!\!\!\! N-C-S \right]_2 Ni$$
$$\|$$
$$S$$

Proprietary names:	Robac NiBUD—Robinson
	Naugard NBC—Uniroyal
	Vanguard N—Vanderbilt
Physical form:	Olive green powder or flake
Acute toxicity:	LD_{50} (oral, rat) greater than 2000 mg kg^{-1}
Skin and eye irritation:	Non-irritant
	May cause dermatitis in persons already sensitised to nickel
Chronic toxicity:	An excess of nasal and lung cancers has been shown to occur in the nickel refining industry
	Nickel dibutyldithiocarbamate has given negative results in one carcinogenicity test on mice[5] and positive results in a second test[6]
	It is not possible with current information to evaluate the potential human carcinogenicity of this compound

COPPER DIMETHYLDITHIOCARBAMATE

$$\begin{array}{c} CH_3 \\ CH_3 \end{array}\!\!\!\! N-C-S-Cu-S-C-N \!\!\!\!\begin{array}{c} CH_3 \\ CH_3 \end{array}$$
$$\|\|$$
$$SS$$

Proprietary names:	Robac CuDD—Robinson
	Cumate—Vanderbilt
Physical form:	Dark brown powder
Acute toxicity:	LD_{50} (oral, rat) greater than 2000 mg kg^{-1}
Skin and eye irritation:	Neither skin nor eye irritant
Chronic toxicity:	Tested for carcinogenicity in mice with negative results[5]

CADMIUM DIETHYLDITHIOCARBAMATE

$$\left[\begin{array}{c} C_2H_5 \\ C_2H_5 \end{array} \!\!\!>\!\! N\!-\!\overset{\overset{S}{\|}}{C}\!-\!S \right]_2 \!\!Cd$$

Proprietary name: Ethyl cadmate—Vanderbilt
Physical form: Solid
Acute toxicity: Maximum tolerated dose $21 \cdot 5 \, mg \, kg^{-1}$
Skin and eye irritation: Non-irritant
Chronic toxicity: Cadmium compounds have been shown to produce a variety of chronic toxic effects:
These include kidney damage, emphysema, skeletal disorders, testicular changes, anaemia and damage to the olfactory organs; some evidence of carcinogenicity has also been reported
Cadmium diethyldithiocarbamate would be likely, if exposure was sufficient, to produce similar toxic effects
An 18-month feeding trial on mice at a dosage of $21 \cdot 5 \, mg \, kg^{-1}$ did not produce an elevation in the tumour incidence[5]

PIPERIDINIUM PENTAMETHYLENE DITHIOCARBAMATE

Proprietary names: Vulkacit P—Bayer
Robac PPD—Robinson
Physical form: Creamy white powder
Acute toxicity: LD_{50} (oral, rat) $450–1000 \, mg \, kg^{-1}$
Skin and eye irritation: Severely irritating to eyes
Skin irritant and sensitiser (man)

Guanides

N,N'-DIPHENYL-GUANIDINE (DPG)

Proprietary names:	Anchor DPG—Anchor
	DPG—Monsanto
	Ekagom D—Ugine Kuhlmann
	Vulcafor DPG—Vulnax
	Vulkacit D—Bayer
Physical form:	White powder or pellets
Acute toxicity:	LD_{50} (oral, rat) 280 mg kg^{-1}
	LD_{50} (dermal, rabbit) 2000–3160 mg kg^{-1}
Skin and eye irritation:	Severely irritating to eyes (man)
	Skin irritant
	Dust irritating when inhaled
	Can cause skin sensitisation (man)

DI-o-TOLYL-GUANIDINE (DOTG)

Proprietary names:	Rhenogran DOTG—Rhein Chemie
	DOTG—Anchor
	Vulcafor DOTG—Vulnax
	Vulkacit DOTG/C—Bayer
Physical form:	White powder
Acute toxicity:	LD_{50} (oral, rat) 134 mg kg^{-1}
Skin and eye irritation:	Severely irritating to eyes (rabbit)
	Not irritating to skin (rabbit)
	Possible skin sensitiser (man)

o-TOLYL-BIGUANIDE (OTBG)

Proprietary name:	Vulkacit 1000/C—Bayer
Physical form:	White powder
Acute toxicity:	LD_{50} (oral, rat) 750–1250 mg kg^{-1}
	LD_{50} (dermal, rabbit) greater than 7940 mg kg^{-1}
Skin and eye irritation:	Severely irritating to eyes (rabbit)

DIORTHOTOLYL-GUANIDINE SALT OF DICATECHOL BORATE

Proprietary name:	Permalux—DuPont
Physical form:	Greyish brown powder
Acute toxicity:	LD_{50} (oral, rat) 570 mg kg^{-1}
Skin and eye irritation:	Skin and eye irritant

Sulphenamides

N-TERT.-BUTYL-2-BENZOTHIAZYL SULPHENAMIDE (TBBS)

Proprietary names:	Delac NS—Uniroyal
	Santocure NS—Monsanto
	Vulkacit NZ—Bayer
Physical form:	Light buff powder or pellets
Acute toxicity:	LD_{50} (oral, rat) 6310 mg kg^{-1}
	LD_{50} (dermal, rabbit) greater than 7940 mg kg^{-1}
Skin and eye irritation:	Neither skin nor eye irritant (rabbit)
	Not a primary skin irritant or sensitiser (human patch tests)

N-CYCLOHEXYL-2-BENZOTHIAZYL SULPHENAMIDE (CBS)

Proprietary names:	Anchor CBS—Anchor
	Delac 5—Uniroyal
	Ekagom CBC—Ugine Kuhlmann
	Santocure—Monsanto
	Vulcafor CBS—Vulnax
	Vulkacit CZ—Bayer
Physical form:	Pale buff coloured powder or pellets
	Sometimes tinted
Acute toxicity:	LD_{50} (oral, rat) 5300 mg kg^{-1}
	LD_{50} (dermal, rabbit) greater than 7940 mg kg^{-1}
Skin and eye irritation:	Not irritating to skin (rabbit)
	Mildly irritating to eyes (rabbit)
	Some clinical evidence of skin sensitisation from contact with finished products but has caused little problem in industrial use
Chronic toxicity:	Tests in mice at a daily dose of 215 mg kg^{-1} over 18 months did not produce an increased number of tumours[5]

N,N-DICYCLOHEXYL-2-BENZOTHIAZYL SULPHENAMIDE (DCBS)

Proprietary names:	Vulcafor DCBS—Vulnax
	Vulkacit BZ/C—Bayer
Physical form:	Pale yellow to slightly pink powder
Acute toxicity:	LD_{50} (oral, rat) 10 000 mg kg^{-1}
Skin and eye irritation:	Not irritating to eyes (rabbit)
	Not a primary irritant or sensitiser (dog)

2-MORPHOLINOTHIOBENZOTHIAZOLE (MBS)

Proprietary names:	Delac MOR—Uniroyal
	NOBS Special—Anchor
	Santocure MOR—Monsanto
	Vulcafor MBS—Vulnax
	Vulkacit MOZ—Bayer
Physical form:	Buff coloured pellets or flakes
	Sometimes tinted
Acute toxicity:	LD_{50} (oral, rat) greater than 7940 mg kg^{-1}
	LD_{50} (dermal, rabbit) greater than 7940 mg kg^{-1}
Skin and eye irritation:	Neither skin nor eye irritant (rabbit)
	May cause skin sensitisation (industrial)
Chronic toxicity:	Tests in mice using daily doses of 464 mg kg^{-1} over 18 months did not produce an increased number of tumours[5]

Short term screening tests carried out by Goodrich gave a mixture of results, some positive and some negative (information from Goodrich, March 1980)

Other Comments
Produces morpholine when heated above 130 °C which is a skin irritant, and may be the cause of the few cases of skin irritation reported during use of this material

Thiazoles

2-MERCAPTOBENZOTHIAZOLE (MBT)

Proprietary names:	Anchor MBT—Anchor
	Captax—Vanderbilt
	Thiotax—Monsanto
	Vulcafor MBT—Vulnax
	Vulkacit Merkapto—Bayer
Physical form:	Yellowish powder, oil treated powder or pellets
	Sometimes tinted
Acute toxicity:	LD_{50} (oral, rat) 3800 mg kg^{-1}
	LD_{50} (dermal, rabbit) greater than 7940 mg kg^{-1}
Skin and eye irritation:	Neither skin nor eye irritant (rabbit)
	Has caused skin sensitisation both in factory use (a limited number) and in persons in contact with rubber articles containing MBT (e.g. footwear)
Chronic toxicity:	Mice fed 100 mg kg^{-1} per day over 18 months did not show an increased number of tumours[5]

2,2'-DIBENZOTHIAZYL DISULPHIDE (MBTS)

Proprietary names:	Altax—Vanderbilt
	Anchor MBTS—Anchor
	Thiofide MBTS—Monsanto
	Vulcafor MBTS—Vulnax
	Vulkacit DM—Bayer
Physical form:	Cream coloured powder or pellets
Acute toxicity:	LD_{50} (oral, rat) greater than 7940 mg kg^{-1}
	LD_{50} (dermal, rabbit) greater than 7940 mg kg^{-1}
Skin and eye irritation:	Neither skin nor eye irritant (rabbit)
Chronic toxicity:	Mice fed 464 mg kg^{-1} per day over 18 months did not show an increased number of tumours[5]
	Some potential mutagenic and teratogenic effects reported[27]

ZINC SALT OF 2-MERCAPTOBENZOTHIAZOLE (ZMBT)

Proprietary names:	Anchor ZMBT—Anchor
	Bantex—Monsanto
	Oxaf—Uniroyal
	Robac MX-1—Robinson
	Vulcafor ZMBT—Vulnax
	Vulkacit ZMBT—Bayer
Physical form:	Cream coloured powder
Acute toxicity:	LD_{50} (oral, rat) 7050 mg kg^{-1}
	LD_{50} (dermal, rabbit) greater than 7940 mg kg^{-1}
Skin and eye irritation:	Neither skin nor eye irritant (rabbit)
	Is likely to give similar skin irritation and sensitisation effects to MBT—see p. 83

2-(2′,4′-DINITROPHENYLTHIO)BENZOTHIAZOLE

Proprietary name:	Ureka base—Monsanto
Physical form:	Yellow powder
Acute toxicity:	LD_{50} (oral, rat) 7130 mg kg^{-1}
	LD_{50} (dermal, rabbit) greater than 7940 mg kg^{-1}
Skin and eye irritation:	Neither skin nor eye irritant (rabbit)

Thioureas

ETHYLENETHIOUREA OR 2-MERCAPTO-IMIDAZOLINE (ETU)

Proprietary names:	Robac 22—Robinson
	Vulkacit NN/C—Bayer
Physical form:	Available as polymer predispersion
Acute toxicity:	LD_{50} (oral, rat) 920 mg kg^{-1}
Skin and eye irritation:	Neither skin nor eye irritant
Chronic toxicity:	Three carcinogenicity studies have given positive results:

    (i) Fed to mice at 215 mg kg^{-1} per day, it gave liver cancer in two different mice strains[5]

    (ii) Fed to rats at 250 ppm of the diet for 12 months it gave a 13% incidence of thyroid tumours[7]

Chronic toxicity: (iii) Fed to rats (26 male and 26 female) for 18 months at two dose levels, it gave results as follows:
At 350 ppm of diet—17 males and 8 females had thyroid cancer
At 175 ppm of diet—3 males and 3 females had thyroid cancer[8]

ETU has also been shown to be a teratogen:
(i) Skin painting tests in pregnant rats:[9]
60 mg kg^{-1}—foetal malformations
40 mg kg^{-1}—no foetal malformations
(ii) Oral feeding tests in pregnant rats and rabbits[10]
Rats:
50 mg kg^{-1}—severe foetal malformations
10 mg kg^{-1}—some foetal malformations
Rabbits:
50 mg kg^{-1}—increase in resorption sites, some decrease in brain weight in offspring
10 mg kg^{-1}—no effects

Other Comments

It has been shown that a number of substances possessing the imidazoline ring may produce teratogenic effects and suggestions have been made that this ring is essential for such an effect in thioureas.[11] DuPont have also shown that positive amounts of ETU can be washed from the surface of polychloroprene products cured with ETU.[9]

Studies of the incidence of thyroid cancers in the rubber industry in one particular region of the UK showed that there was no excess when compared with the national rate.[12] The British Rubber Manufacturers' Association's epidemiological study also showed that no significant excess of thyroid cancer exists in the UK rubber industry.[13] A study of 1929 workers who had used ETU revealed no cases of thyroid cancer.[14]

Finally a study of females who had worked with rubber compounds cured with ETU showed no excess of foetal abnormalities.[14]

The British Rubber Manufacturers' Association recommends[15] that ETU should only be used in non-dusting (pre-dispersed) forms. The highest possible standards of industrial hygiene should be used at all stages of production, including efficient local exhaust ventilation to remove fumes. Women of child bearing age should not be employed on operations involving the handling of ETU or where they may be exposed to fumes containing ETU (e.g. weighing, mixing, calendering, extruding and curing operations). However, handling of cold cured rubber parts containing ETU (e.g. in inspection or packing operations) may be carried out by women with no risk.

1,3-DIETHYLTHIOUREA (DETU)

$$CH_3CH_2NH-\underset{\underset{S}{\|}}{C}-NHCH_2CH_3$$

Proprietary names:	Robac DETU 90 ⎫ DETU PM ⎬ Robinson DETU flake ⎭
Physical form:	Oiled powder, flakes or polymer masterbatch
Acute toxicity:	LD_{50} (oral, rat) 316 mg kg^{-1}
Skin and eye irritation:	Breaks down at curing temperature (180 °C) to give ethyl isothiocyanate and diethylamine; traces of ethylisothiocyanate may be present in the raw materials, in the fumes from curing and in the finished product; this material may cause keratitis of the eyes with temporary scars occurring on the cornea, and this effect has been seen in workers manufacturing rubber strip by high temperature continuous vulcanisation[16] Workers handling car door seals cured with DETU have suffered from dermatitis[17]
Chronic toxicity:	This material is active on the thyroid gland The goitrogenic dose in humans is 200–600 mg per day[18] Carcinogenicity tests: Rats fed 250 ppm of their diet produced cancers[19] Mice fed 500 ppm of their diet did not produce cancers[19]

1,3-DIPHENYL-2-THIOUREA (OR THIOCARBANILIDE) (DPTU)

$$\text{C}_6\text{H}_5\text{—NH—C(=S)—NH—C}_6\text{H}_5$$

Proprietary names:	A1—Monsanto
	Rhenocure CA—Bayer
Physical form:	Off-white or yellowish powder
Acute toxicity:	LD_{50} (oral, rat) 4500 mg kg^{-1}
	LD_{50} (dermal, rabbit) greater than 7940 mg kg^{-1}
Skin and eye irritation:	Neither skin nor eye irritant
	No information is available on the breakdown products at high temperature, but see p. 87
Chronic toxicity:	Rats fed 1000 ppm of their diet for 2 years did not develop cancers[20]
	No reports are available on teratogenicity but see ETU, p. 88

1,1,3-TRIBUTYLTHIOUREA (TBTU)

$$(\text{C}_4\text{H}_9)_2\text{N—C(=S)—N(H)—C}_4\text{H}_9$$

Proprietary name:	Robac Tributu—Robinson
Physical form:	Pale brownish-yellow viscous liquid, boiling point 220–230 °C
Acute toxicity:	LD_{50} (oral, rat) 5500 mg kg^{-1}
	LD_{50} (dermal, rabbit) greater than 10 000 mg kg^{-1}
Skin and eye irritation:	Slightly irritating to skin (rabbits)
	Slightly irritating to eyes (rabbits)
	Breaks down at curing temperature to give butyl isothiocyanate and dibutylamine which makes the fumes irritant to the eyes, nose and throat

Thiurams

TETRAMETHYLTHIURAM MONOSULPHIDE (TMTM)

$$\begin{array}{c} CH_3 \\ \diagdown \\ CH_3 \end{array} N-\underset{\underset{S}{\|}}{C}-S-\underset{\underset{S}{\|}}{C}-N \begin{array}{c} \diagup CH_3 \\ \\ CH_3 \end{array}$$

Proprietary names:	Anchor TMTM—Anchor
	Monex—Uniroyal
	Monothiurad—Monsanto
	Robac TMS—Robinson
	Vulcafor TMTM—Vulnax
	Vulkacit Thiuram MS—Bayer
Physical form:	Yellow powder or pellets
Acute toxicity:	LD_{50} (oral, rat) 1250–1400 mg kg^{-1}
	LD_{50} (dermal, rabbit) greater than 2000 mg kg^{-1}
	Causes unpleasant side effects (e.g. nausea, vomiting, flushing) when both this material and alcohol are present in the body—the 'Antabuse' effect (see TETD, p. 91); inhalation of small quantities may be sufficient to produce this effect, although the effect may be due to TMTD impurity in the commercial material
Skin and eye irritation:	Neither skin nor eye irritant (rabbit)
	Some evidence that it causes skin sensitisation (human patch tests) but cases of industrial sensitisation are uncommon
Chronic toxicity:	Non-carcinogenic in mice[5]

(Preceding entry continuation:)

	Similar isothiocyanates have given temporary scars of the cornea (see DETU, p. 87)
Chronic toxicity:	No information on carcinogenicity or teratogenicity (but see ETU, p. 88)

TETRAMETHYLTHIURAM DISULPHIDE (TMTD)

$$\begin{array}{c} CH_3 \\ \diagdown \\ N-C-S-S-C-N \\ \diagup \| \| \diagdown \\ CH_3 S S CH_3 \end{array} \begin{array}{c} CH_3 \\ \\ \\ CH_3 \end{array}$$

Proprietary names:	Anchor TMTD—Anchor
	Robac TMT—Robinson
	Thiurad—Monsanto
	Thiuram M—DuPont
	Vulcafor TMTD—Vulnax
	Vulkacit Thiuram—Bayer
Physical form:	White powder, granules or predispersion
Acute toxicity:	LD_{50} (oral, rat) 780–1300 mg kg^{-1}
	LD_{50} (dermal, rabbit) greater than 7940 mg kg^{-1}
	Consumption of alcohol after working in an atmosphere containing TMTD may cause unpleasant symptoms, for instance nausea, vomiting, flushing (the 'Antabuse' effect, see TETD, p. 91)
Skin and eye irritation:	Not irritating to eyes (rabbit)
	Prolonged exposure to TMTD caused lachrymation and photophobia in man; these symptoms slowly reversed after removal from exposure; chronic conjunctivitis occurred in 14% of this group[21]
	Causes both primary irritation and sensitisation of the skin (industrial reports)
Chronic toxicity:	Caused neurotoxic effects in 8 out of 24 female rats and 0 out of 24 male rats fed 70 mg per day[22]
	Carcinogenicity:
	Mice fed 10 mg kg^{-1} for 18 months did not show an excess of tumours[5]
	Teratogenicity:
	Positive results in mice[23,24] and hamsters[25] fed 250 mg kg^{-1}; these dose levels are

high, and are likely to be out of the range encountered in industrial use

TETRAETHYLTHIURAM DISULPHIDE (TETD)

$$\begin{array}{c} C_2H_5 \\ \diagdown \\ N\!-\!\!C\!-\!\!S\!-\!\!S\!-\!\!C\!-\!\!N \\ \diagup \| \| \diagdown \\ C_2H_5 S S C_2H_5 \\ \diagup \\ C_2H_5 \end{array}$$

Proprietary names:	Robac TET—Robinson
	Ethyl Tuex—Uniroyal
Physical form:	White powder, flakes, polymer dispersion
Acute toxicity:	LD_{50} (oral, rat) greater than 2000 mg kg^{-1}
	Has been used medically (as the drug 'Antabuse') to treat alcoholism; when TETD and alcohol are present together in the body, unpleasant symptoms, such as nausea, vomiting, flushing and tachycardia, are produced; alcohol is normally metabolised by the body to acetic acid via acetaldehyde; TETD metabolises to give diethylamine and carbon disulphide, and these (particularly CS_2) block the oxidation process for alcohol at the acetaldehyde stage; it is the presence of acetaldehyde in the blood which causes the symptoms described above; the 'Antabuse' effect may occur in workers who consume alcohol after exposure to TETD, particularly as the dust
	During testing of TETD for drug use, it has been shown that single doses of 3–6 g and doses of 0·25–1·0 g over a period of 1 month gave no effects in humans
Skin and eye irritation:	Not irritating to eyes (rabbit)
	Slight primary skin irritant and sensitiser (industrial reports)
Chronic toxicity:	Some evidence of neurotoxicity[26]

TETRABUTYLTHIURAM DISULPHIDE (TBUT)

$$\begin{array}{c} C_4H_9 \\ \diagdown \\ \diagup \\ C_4H_9 \end{array} N-\underset{\underset{S}{\|}}{C}-S-S-\underset{\underset{S}{\|}}{C}-N \begin{array}{c} C_4H_9 \\ \diagup \\ \diagdown \\ C_4H_9 \end{array}$$

Proprietary name:	Robac TBUT—Robinson
Physical form:	Brown, oily, viscous liquid, also available as pellets
Acute toxicity:	LD_{50} (oral, rat) greater than 2000 mg kg^{-1}
	May give 'Antabuse' effect if alcohol is consumed after exposure (see TETD above) but this effect is less likely with TBUT than with TETD or TMTD
Skin and eye irritation:	Neither skin nor eye irritant (rabbit)

DIPENTAMETHYLENETHIURAM DISULPHIDE

Proprietary name:	Robac PTD—Robinson
Physical form:	Creamy white powder
Acute toxicity:	LD_{50} (oral, rat) greater than 500 mg kg^{-1}
Skin and eye irritation:	Slightly irritating to eyes
	Slightly irritating to skin
	Skin sensitiser

DIPENTAMETHYLENETHIURAM TETRASULPHIDE/HEXASULPHIDE

$n = 4$ or 6

Proprietary name: Robac Thiuram P25—Robinson
Physical form: Buff powder, granules or predispersion
Acute toxicity: LD_{50} (oral, rat) 500 mg kg^{-1}
Skin and eye irritation: Skin and eye irritant (industrial reports)

Xanthates

ZINC ISOPROPYL XANTHATE (ZIX)

$$\left[\begin{array}{c} CH_3 \\ CH_3 \end{array} \!\!> CH-O-\underset{\underset{S}{\|}}{C}-S \right]_2 Zn$$

Proprietary name: Robac ZIX—Robinson
Physical form: Creamy white powder
Acute toxicity: LD_{50} (oral, rat) greater than 2000 mg kg^{-1}
Skin and eye irritation: Neither skin nor eye irritant
May cause inflammation of mucuous membranes

Thiazolidine

3-METHYL-2-THIONE-THIAZOLIDINE

$$CH_3-\underset{\underset{CH_2-CH_2}{|}}{N}\!\!\diagup\!\!\overset{\overset{S}{\|}}{C}\!\!\diagdown\!\! S$$

Proprietary name: Vulkacit CRV—Bayer
Physical form: Whitish brown pellets
Acute toxicity: LD_{50} (oral, rat) 1200 mg kg^{-1}
Skin and eye irritation: Mildly irritating to eyes
Not irritating to skin

Miscellaneous

ZINC o,o-DI-n-BUTYLPHOSPHORODITHIOATE

$$\left[\begin{array}{c} C_4H_9O \\ C_4H_9O \end{array} \!\!\!\!> \!\! P \!\! \begin{array}{c} S \\ \| \\ \end{array} \!\! -S \right]_2 \!\! Zn$$

Proprietary name: Vocol—Monsanto
Physical form: Liquid, also available as solid on inert base (Vocol S)
Acute toxicity: LD_{50} (oral, rat) 1800 mg kg^{-1}
LD_{50} (dermal, rabbit) 5010–7940 mg kg^{-1}
Skin and eye irritation: Severely irritating to eyes (rabbit)
Moderately irritating to skin (rabbit)

REFERENCES

1. Clayton, G. D. and Clayton, F. E. (Eds), *Patty's Industrial Hygiene and Toxicology*, Vol. 2A, 3rd edn, Wiley-Interscience, New York, 1981, pp. 2696–700.
2. Della Porta, G., Colnaghi, M. I. and Parmiani, G., Non-carcinogenicity of hexamethylenetetramine in rats and mice, *Arch. Ind. Hyg. & Occ. Med.* (1951), **4**, 119–22.
3. Andrianova, M. M. and Aleikseev, I. V., Carcinogenic properties of Severe, Maneb and Zineb, *Voprise Pitanyia (Moscow)* (1970), **29**, 71–4.
4. Hodge, H. C., Maynard, E. A., Downs, W., Blanchet, H. J. and Jones, C. K., Acute and short term oral toxicity tests of Firbam and Ziram, *J. Am. Pharm. Assoc., Scientific Ed.* (1952), **41**, 662–5.
5. Innes, J. R. M., *et al.*, Bioassay of pesticides and industrial chemicals for tumourgenicity in mice, a preliminary note, *J. Nat. Cancer Inst.* (1969), **42**, 1101–14.
6. *National Technical Information Service PB Report 223, 159*, USA, 1976.
7. *Food Chemical News*, 20 September 1971, p. 18 (also quoted in ref. 9).
8. Ulland, B. M., Weisburgar, J. H., Weisburgar, E., Rice, J. M. and Cypher, E., Thyroid cancer in rats from ethylenethiourea intake, *J. Nat. Cancer Inst.* (1972), **49**, 583.
9. *NA-22 Handling Precautions & Toxicity, Data Sheet 14A*, E. I. DuPont de Nemours & Co., May 1972.
10. Khera, K. S., The teratogenicity of ethylene thiourea, *Teratology* (1973), **7**, 243.

11. Ruddick, J. A., Newsome, W. H. and Nash, L., Correlation of teratogenicity and molecular structure: ethylene thiourea and related compounds, *Teratology* (1976), **13**, 263.
12. Parkes, H. G., Living with carcinogens, *J. Inst. Rubber Ind.* (1974), **8**, 21–3.
13. Parkes, H. G., Veys, C. A., Waterhouse, J. A. H. and Peters, A., Cancer mortality in the British rubber industry, *Brit. J. Indust. Med.* (1982), **39**, 209–20.
14. Smith, D., Ethylenethiourea—a study of possible teratogenicity and thyroid carcinogenicity, *J. Soc. Occ. Med.* (1976), **26**, 72–4.
15. *Ethylene Thiourea*, Bulletins No. 11 and 16, British Rubber Manufacturers' Association.
16. Groves, J. S. and Smail, J. M., Outbreak of superficial keratitis in rubber workers, *Brit. J. Ophthalmology* (1959), **53**, 683–7.
17. White, W. G. and Vickers, H. R., Diethylthiourea as a cause of dermatitis in a car factory, *Brit. J. Indust. Med.* (1970), **27**, 167–9.
18. Williams, R. H., Antithyroid drugs, 1. Tetramethyl thiourea and diethylthiourea, *J. Clin. Endocrinology* (1945), **5**, 210–16.
19. *Public Bulletin Report 288-626*, National Cancer Institute, USA.
20. Mohr, H. J. and Northduft, H., *Int. Arch. für Gewerbepathologie und Gewerdehygiene* (1967), **23**, 168–74.
21. Sivitskaya, I. I., State of the organ of vision in persons working in contact with TMTD, *Optalmologisceskii Zeilung* (1974), **28**, 286–8.
22. Lee, C. C. and Peters, P. J., Neurotoxicity and behaviour effects of thiuram in rats, *Envir. Health Perspectives* (1967), **17**, 35–43.
23. Roll, R., Teratologische Untersuchungen mit Thiram (TMT) an Zwei Mauseslammen, *Arch. Toxikology* (1971), **27**, 173–86.
24. Matthiaschk, G., Über der Einfluss von L-Cystein auf der Teratogenese durch Thiram (TMTD) bei NMRI-Mausen, *Arch. Toxicology* (1973), **30**, 251–63.
25. Robens, J. F., Teratologic studies on carbaryl diazinon, norea, disulfiram and thiram in small laboratory animals, *Toxicology & Appl. Pharmacol.* (1969), **15**, 156–63.
26. Poitou, P., Marignac, B. and Gradisky, D., Effect of tetramethylthiuram monosulphide and disulphide on the central nervous system. Comparison with disulfiram, *J. Pharmacol. (Paris)* (1978), **9**, 35–44.
27. Plastics chemical causes mutagenic, teratogenic, embryotoxic effects, EPA says, *Chemical Regulator Reporter* (1983), **7** (22), 736.

9

Retarders

N-NITROSODIPHENYLAMINE

$$\text{Ph–N(Ph)–N=O}$$

Proprietary names:	Curetard A—Monsanto
	Retarder J—Uniroyal
	Vulcatard A—Vulnax
	Vulkalent A—Bayer
Physical form:	Brown granules
Acute toxicity:	LD_{50} (oral, rat) greater than 2000 mg kg^{-1}
Skin and eye irritation:	Neither skin nor eye irritant
Chronic toxicity:	Negative results for carcinogenicity in two series of tests[1,2]
	Weak positive results in one series of tests on mice[3]
	Acts as a nitrosating agent in formulations which produce secondary amines, converting these to the corresponding N-nitroso compounds which have been shown to be present in the atmosphere of curing and other hot rubber processes; some of these materials are carcinogenic (see Part I, Chapter 3, pp. 44–8); for this reason this retarder has now largely been withdrawn from use by the suppliers

RETARDERS

CYCLOHEXYLTHIOPHTHALIMIDE

Proprietary name:	Santogard PVI—Monsanto
Physical form:	Light tan dust suppressed powder or granules containing 25% and 50% active ingredients
Acute toxicity:	LD_{50} (oral, rat) 2600 mg kg^{-1}
	LD_{50} (dermal, rabbit) greater than 5010 mg kg^{-1}
Skin and eye irritation:	Neither skin nor eye irritant (rabbit) but patch tests in humans have shown it to be a primary skin irritant and sensitiser
	Patch tests on human subjects using a rubber stock with 2 phr produced negative results for primary skin irritation and sensitisation
	Eye irritant (industrial use)

Other Comments
The crystals which may form in the vicinity of heated rubber operations containing PVI consist of sublimed phthalimide.

SALICYLIC ACID

Proprietary names:	Vulcatard SA—Vulnax
	Retarder TSA—Monsanto
Physical form:	White oil treated powder
Acute toxicity:	LD_{50} (oral, rat) 890 mg kg^{-1}
	LD_{50} (dermal, rabbit) 3160–5010 mg kg^{-1}
Skin and eye irritation:	Moderately irritating to eyes (rabbit)
	Slightly irritating to skin (rabbit)

PHTHALIC ANHYDRIDE

Proprietary names:	Phthalic anhydride—ICI
	Retarder ESEN—Uniroyal
	Retarder PD—Anchor
Physical form:	White to pale cream powder or flake
Acute toxicity:	LD_{50} (oral, rat) 800–1600 mg kg^{-1}
Skin and eye irritation:	Severely irritating to skin, eyes and respiratory system
	Skin sensitiser (industrial use)
Chronic toxicity:	May cause respiratory sensitisation[4-6]

REFERENCES

1. Druckrey, H., et al., Zeitschr. f. Krebsforschg. (1967), **69**, 100.
2. Boyland, E., et al., European J. Cancer (1968), **4**, 233–9.
3. National Technical Information Service PB 223, 159, USA, 1976.
4. Clayton, G. D. and Clayton, F. E. (Eds), Patty's Industrial Hygiene and Toxicology, Vol. 2, 2nd Edn, Interscience, New York, p. 1822.
5. Baader, E. W., Arch. Gewerbepathol. (1955), **13**, 419.
6. Menschick, H., Arch. Gewerbepathol. (1955), **13**, 454.

10

Antidegradants

Condensation Products of Aldehydes and Ketones with Amines and Alcohols

ACETONE–DIPHENYLAMINE CONDENSATION PRODUCT (ADPA)

This material has a complex structure which cannot be represented by a simple chemical formula.

Proprietary names: *Liquid forms:*
BLE 25—Uniroyal
Permanax BL, BLN—Vulnax
Solid forms:
Anchor ADPA (rods)—Anchor
Permanax B (pastilles)—Vulnax
Permanax BLW (on inert carrier)—Vulnax

Physical form: Dark reddish-brown, viscous liquid
Dark brown rods or powder

Acute toxicity: LD_{50} (oral, rat) 2270 mg kg^{-1}

Skin and eye irritation: Neither skin nor eye irritant (rabbit)

Other Comments
Commercial diphenylamine, which is a starting product for this material, contains traces of the potent bladder carcinogen 4-aminodiphenyl, and it has been suggested that this material could be found in the finished product. However, tests have shown that commercial ADPA contains no detectable 4-aminodiphenyl (less than 1 ppm) and it is likely that any traces of this material in the starting product react with acetone and are thus removed from the finished product (see Part I, Chapter 3, p. 34).

XYLENOL–ALDEHYDE CONDENSATION PRODUCT

$R = CH_2$ or C_4H_8

Proprietary names: Permanax EXP—Vulnax
Physical form: Dark brown resinous beads
Acute toxicity: LD_{50} (oral, rat) greater than $2000\,mg\,kg^{-1}$
Skin and eye irritation: Slightly irritating to skin and eyes

Quinolines

POLYMERISED 2,2,4-TRIMETHYL-1,2-DIHYDROQUINOLINE (TMQ)

Proprietary names: Anchor TMQ (powder/flakes)—Anchor
Flectol (flakes or pastilles)—Monsanto
Permanax TQ—Vulnax
Vulkanox HS—Bayer
Lorvinox ACP—Chemox
Naugard Q—Uniroyal
Physical form: Light brown powder, flakes, pellets or pastilles
Acute toxicity: LD_{50} (oral, rat) $2250\,mg\,kg^{-1}$
LD_{50} (dermal, rabbit) greater than $5010\,mg\,kg^{-1}$
Skin and eye irritation: Neither skin nor eye irritant (rabbit)
No irritation or sensitisation reported (industrial use)

6-ETHOXY-2,2,4-TRIMETHYL-1,2-DIHYDROQUINOLINE (ETMQ)

Proprietary name:	Santoflex AW—Monsanto
Physical name:	Dark brown viscous liquid
Acute toxicity:	LD_{50} (oral, rat) 4400 mg kg^{-1}
	LD_{50} (dermal, rabbit) 5010–7940 mg kg^{-1}
	The vapour generated by allowing Santoflex AW to contact a metal surface heated to 162.5 °C caused mild lachrymation but no mortalities of rabbits exposed to the atmosphere for 5–6 h per day on 3 consecutive days (concentration of Santoflex AW in the atmosphere, 0.195 mg litre^{-1})
Skin and eye irritation:	Not irritating to eyes (rabbit)
	Slightly irritating to skin (rabbit)

Substituted Naphthylamines

PHENYL α-NAPHTHYLAMINE (PAN)

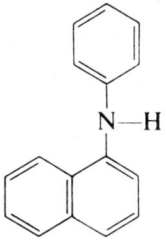

Proprietary names:	Vulkanox PAN—Bayer
	Altofane A—Ugine Kuhlman
Physical form:	Brownish purple powder or pellets
Acute toxicity:	LD_{50} (oral, rat) 1625 mg kg^{-1} (NIOSH)
Skin and eye irritation:	A few cases of skin irritation have been reported (industrial use)

Other Comments
May contain small traces of β-naphthylamine, a human bladder carcinogen (see Part I, Chapter 3).

PHENYL β-NAPHTHYLAMINE (PBN)

Proprietary names:	Anchor PBN (powder)—Anchor
	Anchor PBN (flakes)—Anchor
	Vulkanox PBN—Bayer
Physical form:	Greyish powder or buff flakes
Acute toxicity:	LD_{50} (oral, rat) 8700 mg kg^{-1}
Skin and eye irritation:	Neither skin nor eye irritant (rabbit)
	A few cases of skin irritation have been reported in industrial use
Chronic toxicity:	No tumours or other toxic effects from dogs fed 540 mg PBN per day, 5 days per week for 4–4½ years[1]

Other Comments
PBN, which has been used in the UK rubber industry since 1928, contained some 30 ppm of the human bladder carcinogen β-naphthylamine up to 1970. Since that date some commercial varieties have contained a reduced amount of β-naphthylamine, below 1 ppm. Besides this residual content of β-naphthylamine it has been discovered that when PBN is ingested by some humans, a small proportion is dephenylated within the body to produce β-naphthylamine which can then be detected in the urine.[2] Only minute quantities of β-naphthylamine are produced in this way, but the information, which was initially discovered in a study of workers on a synthetic rubber production facility where PBN was added to the polymer as a stabiliser, was sufficient to cause the manufacturers to replace PBN by an alternative stabiliser. This metabolic dephenylation has also been demonstrated in dogs but the carcinogenic metabolite of β-naphthylamine, 2-naphthylhydroxylamine, could not be demonstrated either in dogs or man. PBN is

now classified as an experimental carcinogen by ACGIH. In spite of these experimentally demonstrated problems, however, several surveys have shown that the rate of bladder cancer in rubber workers who have used PBN is no higher than in the general population. (See Chapters 2 and 3.)

para-Phenylenediamines

N,N'-BIS(1,4-DIMETHYLPENTYL)-*p*-PHENYLENEDIAMINE (77PD)

$$\text{CH}_3\text{-CH(CH}_3\text{)-CH}_2\text{-CH}_2\text{-CH(CH}_3\text{)-NH-C}_6\text{H}_4\text{-NH-CH(CH}_3\text{)-CH}_2\text{-CH}_2\text{-CH(CH}_3\text{)-CH}_3$$

Proprietary names:	Flexone 4L—Uniroyal
	Santoflex 77—Monsanto
	UOP 788—Universal Mathey Products
	Vulkanox 4030—Bayer
Physical form:	Reddish brown viscous liquid
Acute toxicity:	LD_{50} (oral, rat) 730 mg kg^{-1}
	LD_{50} (dermal, rabbit) 3160–5010 mg kg^{-1}
Skin and eye irritation:	Not irritating to eyes (rabbit)
	Slightly irritating to skin (industrial reports)
	Not a skin sensitiser in human patch tests

N,N'-BIS(1-METHYLHEPTYL)-*p*-PHENYLENEDIAMINE

$$\text{CH}_3\text{-(CH}_2)_5\text{-CH(CH}_3\text{)-NH-C}_6\text{H}_4\text{-NH-CH(CH}_3\text{)-(CH}_2)_5\text{-CH}_3$$

Proprietary name:	UOP 288—Universal Mathey Products
Physical form:	Dark brown/red liquid
Acute toxicity:	LD_{50} (oral, rat) 2400 mg kg^{-1}
	LD_{50} (dermal, rabbit) 1800 mg kg^{-1}
Skin and eye irritation:	Mild to moderate eye irritation (rabbit)
	Potential skin sensitiser (industrial)

N,N'-BIS(1-ETHYL-3-METHYLPENTYL)-p-PHENYLENEDIAMINE (DOPD)

```
        CH₃         C₂H₅
         |           |
CH₃—CH₂—CH—CH₂—CH
                      \
                       \
                   C₂H₅       CH₃
                    |          |
     —NH—⟨⟩—NH—CH—CH₂—CH—CH₂—CH₃
```

Proprietary names: UOP 88—Universal Mathey Products
Flexone 8L—Uniroyal
Acute toxicity: LD_{50} (oral, rat), 2100 mg kg^{-1}
LD_{50} (dermal, rabbit) 5010–7940 mg kg^{-1}
Skin and eye irritation: Not irritating to eyes (rabbit)
Slightly irritating to skin (rabbit)
Reported skin sensitiser (industrial)

N,N'-DICYCLOHEXYL-p-PHENYLENEDIAMINE

⟨⟩—NH—⟨⟩—NH—⟨⟩

Proprietary names: UOP 26—Universal Mathey Products
Flexone 18F—Uniroyal
Physical form: Brown or purple flakes
Acute toxicity: LD_{50} (oral, rat) 1470 mg kg^{-1}
Skin and eye irritation: Neither skin nor eye irritant (rabbit)
Some cases of dermatitis reported (industrial)

N-ISOPROPYL-N'-PHENYL-p-PHENYLENEDIAMINE

```
  CH₃
   |
   CH—NH—⟨⟩—NH—⟨⟩
   |
  CH₃
```

Proprietary names: Anchor IPPD—Anchor
Flexone 3C—Uniroyal
Santoflex IP—Monsanto
Permanax IPPD—Vulnax
Vulkanox 4010NA—Bayer

Physical form:	Dark purplish brown flakes
Acute toxicity:	LD_{50} (oral, rat) 900 mg kg^{-1}
	LD_{50} (dermal, rabbit) greater than 7940 mg kg^{-1}
Skin and eye irritation:	Not irritating to eyes (rabbit)
	Has caused a significant incidence of skin sensitisation in industrial use

N-PHENYL-N'-1-METHYLHEPTYL-p-PHENYLENEDIAMINE

$$\text{C}_6\text{H}_5\text{—NH—C}_6\text{H}_4\text{—NH—CH(CH}_3\text{)—(CH}_2)_5\text{—CH}_3$$

Proprietary name:	UOP 688—Universal Mathey Products
Physical form:	Brown viscous liquid
Acute toxicity:	LD_{50} (oral, rat) greater than 2000 mg kg^{-1}
Skin and eye irritation:	Neither skin nor eye irritant (rabbit)
	Some cases of dermatitis reported (industrial)

N-1,3-DIMETHYLBUTYL-N'-PHENYL-p-PHENYLENEDIAMINE

$$(\text{CH}_3)_2\text{CH—CH}_2\text{—CH(CH}_3)\text{—NH—C}_6\text{H}_4\text{—NH—C}_6\text{H}_5$$

Proprietary names:	Santoflex 13—Monsanto
	Permanax 6PPD—Vulnax
	Flexone 7L—Uniroyal
	UOP 588—Universal Mathey Products
	Vulkanox 4020—Bayer
Physical form:	Dark crystals or flakes
Acute toxicity:	LD_{50} (oral, rat) 3580 mg kg^{-1}
	LD_{50} (dermal, rabbit) greater than 7940 mg kg^{-1}

Skin and eye irritation: Not irritating to eyes (rabbit)
Non-primary skin irritant (rabbit)
Experimental skin sensitiser (man); so far, there is little reported evidence of skin sensitisation in industrial use

N,N'-DIPHENYL-p-PHENYLENEDIAMINE

Proprietary names: Permanax DPPD—Vulnax
Anchor DPPD—Anchor
J-2-F—Uniroyal
Physical form: Light grey to black powder, rods
Acute toxicity: LD_{50} (oral, rat) greater than 10000 mg kg^{-1}
Skin and eye irritation: Not irritating to eyes (rabbit)
Not irritating to skin (man)
Some evidence of skin sensitisation potential
Chronic toxicity: Mice fed 10 mg kg^{-1} for 18 months did not show an increased tumour incidence[3]
Has produced adverse effects when fed to pregnant rats (an increase in the gestation period and in the number of offspring born dead);[4] this fact has caused its use as a poultry food additive to be abandoned

N-PHENYL-N'-(p-TOLUENESULPHONYL)-p-PHENYLENEDIAMINE

Proprietary name: Aranox—Uniroyal
Physical form: Brown flake
Acute toxicity: LD_{50} (oral, rat) greater than 2000 mg kg^{-1}
Skin and eye irritation: No skin irritation or sensitisation reported (industrial)

N,N'-DI-β-NAPHTHYL-p-PHENYLENEDIAMINE (DNPD)

Proprietary name: Anchor DNPD—Anchor
Physical form: Light grey powder
Acute toxicity: LD_{50} (oral, rat) 4500 mg kg^{-1}
Skin and eye irritation: Slightly irritating to skin
Skin sensitiser (man)

Other Comments
Contains around 50 ppm of the human bladder carcinogen β-naphthylamine; however, experimental work has indicated that no metabolic conversion to β-naphthylamine occurs with this material (see Part I, Chapter 3, pp. 33–4)

MIXED DIARYL-p-PHENYLENEDIAMINES

This mixture cannot be represented by a single chemical formula.

Proprietary names: Wingstay 100—Goodyear
Vulkanox 3100—Bayer
Physical form: Dark brown flakes
Acute toxicity: LD_{50} (oral, rat) greater than 2000 mg kg^{-1}
Skin and eye irritation: Not irritating to eyes (rabbit)
No skin irritation or sensitisation reported (industrial use)

Diphenylamines

4,4′-DIHEPTYLDIPHENYLAMINE

$$C_7H_{15}-\text{C}_6H_4-NH-\text{C}_6H_4-C_7H_{15}$$

Proprietary names:	Permanax HD—Vulnax
	HDPA—Anchor
Physical form:	Dark brown, viscous liquid
Acute toxicity:	LD_{50} (oral, rat) greater than 2000 mg kg^{-1}
Skin and eye irritation:	Not irritating to skin
	Mildly irritating to eyes

OCTYLATED DIPHENYLAMINE

$$(C_8H_{17})_n-\text{C}_6H_4-NH-\text{C}_6H_4-(C_8H_{17})_n \quad n = 1 \text{ or } 2$$

Proprietary names:	Permanax OD—Vulnax
	Octamine—Uniroyal
Physical form:	Light brown, waxy granules
Acute toxicity:	LD_{50} (oral, rat) greater than 11 000 mg kg^{-1}
Skin and eye irritation:	Not irritating to skin (guinea pig)
	Slightly irritating to eyes (rabbit)
	No reported skin or eye irritation (industrial)

NONYLATED DIPHENYLAMINE

$$(C_9H_{19})_n-\text{C}_6H_4-NH-\text{C}_6H_4-(C_9H_{19})_n \quad n = 1 \text{ or } 2$$

Proprietary name:	Polylite—Uniroyal
Physical form:	Dark brown, viscous liquid
Acute toxicity:	LD_{50} not available
Skin and eye irritation:	Slightly irritating to skin
	Mildly irritating to eyes

DIPHENYLAMINE/α-METHYLSTYRENE REACTION PRODUCT

$$\left[\begin{array}{c}CH_3\\|\\\bigcirc-C-\\|\\CH_3\end{array}\right]_n \bigcirc-NH-\bigcirc\left[\begin{array}{c}CH_3\\|\\-C-\bigcirc\\|\\CH_3\end{array}\right]_n \quad n = 1 \text{ or } 2$$

Proprietary name: Naugard 445—Uniroyal
Physical form: White to brown flake or crystals
Acute toxicity: LD_{50} (oral, rat) greater than 2000 mg kg^{-1}
Skin and eye irritation: Neither skin nor eye irritant

Phenols

STYRENATED PHENOLS

$$\begin{array}{c}OH\\|\\\bigcirc-\left[\begin{array}{c}H\\|\\-C-\bigcirc\\|\\CH_3\end{array}\right]_n\end{array}$$

Proprietary names: Anchor SPH—Anchor
Naugard SP—Uniroyal
Montaclere—Monsanto
Permanax SP(L)—Vulnax
Wingstay S—Goodyear
Physical form: Light yellowish, transparent, viscous liquid
Acute toxicity: LD_{50} (oral, rat) 3550 mg kg^{-1}
LD_{50} (dermal, rabbit) greater than 7940 mg kg^{-1}
Male rats exposed to inhalation of 0·21 mg litre^{-1} styrenated phenol vapour for 6 h showed no observable effects over a 14 day observation period[5]
Skin and eye irritation: No skin or eye irritation (rabbit) or sensitisation (industrial use) reported

Hindered Phenols

2,6-DITERT.-BUTYL-p-CRESOL (BHT)

Proprietary names:	Naugard BHT—Uniroyal Vulkanox KB—Bayer
Physical form:	White crystals
Acute toxicity:	LD_{50} (oral, rat) 1700–1970 mg kg^{-1}
Skin and eye irritation:	No skin or eye irritation reported (industrial use)
Chronic toxicity:	Mice fed BHT as 0·75% of their diet for 16 months gave a 64% incidence of lung tumours as compared to 24% in control animals[6] BHT is used as an antioxidant for food and should cause no carcinogenicity hazard during conventional industrial use

2,4-DIMETHYL-6-(1-METHYLCYCLOHEXYL)PHENOL

Proprietary name:	Permanax WSL—Vulnax
Physical form:	Pale straw coloured, mobile liquid (a solid form on an inert carrier is also available)
Acute toxicity:	LD_{50} (oral, rat) greater than 2000 mg kg^{-1}
Skin and eye irritation:	Severely irritating to skin and eyes

DISTYRENATED HYDROXYTOLUENE OR DISTYRENATED *p*-CRESOL

$$\text{HO-C}_6\text{H}_3(\text{CH}_3)-[\text{CH}(\text{CH}_3)-\text{C}_6\text{H}_5]_2$$

Proprietary name:	Naugard 431—Uniroyal
Physical form:	Amber liquid
Acute toxicity:	LD_{50} (oral, rat) greater than 2000 mg kg^{-1}
Skin and eye irritation:	Neither skin nor eye irritant

2,2′-METHYLENE-BIS(4-METHYL-6-TERT.-BUTYLPHENOL)

$$(\text{CH}_3)_3\text{C}-\text{C}_6\text{H}_2(\text{OH})(\text{CH}_3)-\text{CH}_2-\text{C}_6\text{H}_2(\text{OH})(\text{CH}_3)-\text{C}(\text{CH}_3)_3$$

Proprietary names:	Antioxidant 2246—Anchor
	Vulkanox BKF—Bayer
Physical form:	Colourless to light cream, crystalline powder
Acute toxicity:	LD_{50} (oral, rat) greater than 10000 mg kg^{-1}
Skin and eye irritation:	Neither skin nor eye irritant

2,2′-METHYLENE-BIS(4-METHYL-6-NONYLPHENOL)

$$\text{C}_9\text{H}_{19}-\text{C}_6\text{H}_2(\text{OH})(\text{CH}_3)-\text{CH}_2-\text{C}_6\text{H}_2(\text{OH})(\text{CH}_3)-\text{C}_9\text{H}_{19}$$

Proprietary name:	Naugawhite—Uniroyal
Physical form:	Viscous, amber liquid

Acute toxicity: LD$_{50}$ (oral, rat) greater than 2000 mg kg^{-1}
Skin and eye irritation: Neither skin nor eye irritant
Chronic toxicity: 90-day test (rats) showed no effects[7]

2,2′-METHYLENE-BIS(6-[1-METHYLCYCLOHEXYL]-p-CRESOL)

Proprietary name: Permanax WSP—Vulnax
Physical form: White to pale cream powder
Acute toxicity: LD$_{50}$ (oral, rat) greater than 2000 mg kg^{-1}
Skin and eye irritation: Not irritating to eyes
Slightly irritating to skin

4,4′-BUTYLIDENE-BIS(6-TERT.-BUTYL-m-CRESOL)

Proprietary name: Santowhite Powder—Monsanto
Physical form: Colourless powder
Acute toxicity: LD$_{50}$ (oral, rat) greater than 7940 mg kg^{-1}
LD$_{50}$ (dermal, rabbit) greater than 7940 mg kg^{-1}
Skin and eye irritation: Neither skin nor eye irritant (rabbit)
Not a primary irritant or sensitiser (human patch tests)

2,2′-NONYLENE-BIS(4,6-DIMETHYLPHENOL)

Proprietary name: Permanax WSO—Vulnax
Physical form: White, crystalline powder
Acute toxicity: LD_{50} (oral, rat) greater than 2000 mg kg^{-1}
Skin and eye irritation: Neither skin nor eye irritant

Sulphides

DI(5-TERT.-BUTYL-4-HYDROXY-2-METHYL-PHENYL)SULPHIDE OR 4,4′-THIOBIS(6-TERT.-BUTYL-m-CRESOL)

Proprietary names: Santowhite Crystals—Monsanto
Lorvinox 44S36—Chemox
Lorvinox 44S36P—Chemox
Physical form: White, light grey or buff powder
Acute toxicity: LD_{50} (oral, rat) 4150 mg kg^{-1}
LD_{50} (dermal, rabbit) 5010–7940 mg kg^{-1}
Skin and eye irritation: Neither skin nor eye irritant (rabbits)
Can cause eye and respiratory irritation in industrial use

REFERENCES

1. *Memorandum to Customers*, E. I. DuPont de Nemours & Co, Haskell Laboratory, Wilkington, Delaware, USA, 18 August 1971.
2. Kummer, R. and Tordoir, W. F., Phenyl-β-naphthylamine (PBNA)—another carcinogenic agent? *T. Soc. Geneesk.* (1975), **53**, 415–19.
3. Innes, J. R. M., *et al.*, Bioassay of pesticides and industrial chemicals for tumourigenicity in mice, a preliminary note, *J. Nat. Cancer Inst.* (1972), **49**, 583.
4. Oser, B. L. and Oser, M., Inhibitory effect of feed grade diphenyl-*p*-phenylenediamine on parturition in rats, *J. Agric. Food Chem.* (1956), **4**, 796–7.
5. *Toxicity Information Sheet on Montaclere 10.10.75, Y-74-118;* Monsanto, Louvain-la-Neuve, Belgium.
6. Clapp, N. K., Tyndall, R. L., Cumming, R. B. and Otten, J., Effects of butylated hydroxytoluene alone or with dimethylnitrosamine in mice, *Food & Cosmetics Toxicol.* (1974), **12**, 367–71.
7. *Safety Information on Naugawhite*, Uniroyal, Naugatuck, USA.

11

Blowing Agents

AZODICARBONAMIDE

$$\begin{array}{c} N\text{—}CO\text{—}NH_2 \\ \| \\ N\text{—}CO\text{—}NH_2 \end{array}$$

Proprietary names:	Celogen AZ—Uniroyal
	Genitron AC—FBC
	Porofor ADC/R—Bayer
	Azobul—Ugine Kuhlmann
Physical form:	Fine yellow powder
Acute toxicity:	LD_{50} (oral, rat) greater than 680 mg kg^{-1}
Skin and eye irritation:	No information
Chronic toxicity:	This is a fine, dusty material, particle size 2–10 μm: exposure to the dust has been shown to produce occupational asthma in 28 out of 151 production workers;[1] respiratory sensitisation in these workers was demonstrated by a repetition and worsening of the symptoms after re-exposure; care should be taken to prevent inhalation of dust from this blowing agent

BENZENESULPHONYL HYDRAZIDE

$SO_2-NH-NH_2$ (phenyl ring)

Proprietary names:	Celogen BSH paste—Uniroyal
	Genitron BSH paste or powder—FBC
	Porofor BSH paste or powder—Bayer
Physical form:	White or cream powder, grey paste
Acute toxicity:	LD_{50} (oral, rat) 112 mg kg^{-1} (paste)
Skin and eye irritation:	Some evidence for skin sensitisation
	Irritating to eyes
	Embryotoxicity in experimental mice at doses of 62 mg kg^{-1} has been reported[2]

DINITROSOPENTAMETHYLENETETRAMINE

$$ON-N \begin{matrix} CH_2-N-CH_2 \\ | \quad\quad | \\ CH_2 \; N-NO \\ | \quad\quad | \\ CH_2-N-CH_2 \end{matrix}$$

Proprietary names:	Porofor DNO—Bayer
	Vulcacel BN 94 (75%)—Vulnax
	Unicel ND (40%)—DuPont
	Dipentax—Ugine Kuhlmann
Physical form:	Cream powder
Acute toxicity:	LD_{50} (oral, rat) 940 mg kg^{-1}
Skin and eye irritation:	Not irritating to skin (rabbit)
	Slightly irritating to eyes (rabbit)
Chronic toxicity:	Shown to have some effects on the central nervous system[3]
	No tumours reported when administered to rats, intraperitoneally[4] or orally[5] in daily doses up to 9 mg; the evidence for its non-carcinogenicity has been reviewed[6]
	A potential generator of other nitrosamines in cure systems producing secondary amines (see Part I, Chapter 3).

p,p'-OXY-BIS(BENZENESULPHONYL HYDRAZIDE) (OBSH)

$$H_2N-NH-SO_2-\underset{}{\bigcirc}-O-\underset{}{\bigcirc}-SO_2-NH-NH_2$$

Proprietary names:	Celogen OT—Uniroyal
	Genitron OB—FBC
Physical form:	Pale cream powder
Acute toxicity:	LD_{50} (oral, rat) 5200 mg kg^{-1}
Skin and eye irritation:	Slightly irritating to skin (industrial use)
Chronic toxicity:	The dust produced is very fine, and care should be taken to prevent inhalation

AZOBIS-ISOBUTYRONITRILE (AZDN)

$$CH_3-\underset{CH_3}{\overset{CN}{C}}-N=N-\underset{CH_3}{\overset{CN}{C}}-CH_3$$

Proprietary names:	Genitron AZDN—FBC
	Vazo 64—DuPont
Physical form:	White, crystalline solid
Acute toxicity:	LD_{50} (oral, rat) 500 mg kg^{-1}
	LD_{50} (dermal, rat) $\gg$ 500 mg kg^{-1}
	Breaks down at temperatures above 26 °C (rapidly above 100 °C) to give tetramethylsuccinonitrile (TMSN):

$$CH_3-\underset{CH_3}{\overset{CN}{C}}-N=N-\underset{CH_3}{\overset{CN}{C}}-CH_3 \longrightarrow$$

$$N_2 + CH_3-\underset{CH_3}{\overset{CN}{C}}-\underset{CH_3}{\overset{CN}{C}}-CH_3$$

Acute toxicity: TMSN is a much more toxic material than AZDN itself, and has produced symptoms of headache, nausea and convulsions in workers making PVC foam products;[7-9] has also been shown to produce convulsions and death in animals at 25 mg kg^{-1} (oral, rat), 60 ppm for 2/3 h or 6 ppm for 30 h (inhalation, rat)[10] and 28 ppm for 3 h or 22 ppm for $3\frac{1}{2}$ h (inhalation, mouse);[11] threshold limit value for TMSN in the atmosphere is 0·5 ppm; during use of AZDN, care should be taken that atmospheric concentrations of TMSN do not exceed 0·5 ppm, and that no significant quantities of TMSN remain in the finished product

Skin and eye irritation: Not irritating to skin
Mildly irritating to eyes

Chronic toxicity: No cumulative effects of TMSN were noted in animal experiments[12]

REFERENCES

1. Slovak, A. J. M., Occupational asthma caused by a plastics blowing agent, azodicarbonamide, *Thorax* (1981), **36**, 906-9.
2. Matschke, G. H. and Fagerstone, K. R., *J. Tox. Envir. Health* (1977), **3**, 407-11.
3. Desi, I., Bardas, S., Lehotzky, K. and Hajman, B., *Medicina del Lavora (Milan)* (1967), **58**, 22-31.
4. Boyland, E., Carter, R. L., Gorrod, J. W. and Roe, F. J. C., *European J. Cancer* (1968), **4**, 233-9.
5. Weisburger, J. H., Weisburger, E. K., Mantal, H., Hadidian, Z. and Frederickson, T., *Naturwissenschaften* (1966), **53**, 508.
6. *IARC Monograph No. 11*, International Agency for Research on Cancer, Lyon, France, 1976, p. 241.
7. *Documentation to the Threshold Limit Values*, American Conference of Governmental Industrial Hygienists, Cincinnati, Ohio, 1982.
8. Reinl, W., *Arch. Toxikol.* (1957), **16**, 367.
9. Quoss, H., *Gesundheitsgefahren in der Kunststoffindustrie*, Joh. A. Barth, Leipzig, 1959.

10. Harger, R. N. and Hulpieu, H. R., Toxicity of tetramethylsuccinonitrile and the antidotal effects of thiosulphate, nitrite and barbiturates, *Federation Proceedings* (*USA*) (1949), **8**, 205. (Quoted in US Bureau of Mines Report No. 4777, 1951.)
11. Hecht, G. and Kimmerle, H., 'Toxicologische Untersuchungen über Tetramethylbernsteinsauredinitril,' unpublished investigations quoted by Reinl, ref. 8.
12. Oettel, H., *Arch. Exptl. Pathol. Pharmakol.* (1958), **232**, 77.

12

Solvents

A wide range of solvents is in use in the rubber industry. Inhalation of excessive concentrations of any of these solvents is likely to produce adverse effects—fatigue, headache, drowsiness and nausea—and may if continued lead to loss of consciousness. These effects can be prevented by ensuring that the solvent concentrations are maintained below the threshold limit value for the solvent or solvents in question. Direct skin contact with these solvents produces defatting of the skin and this can lead to further skin problems if contact continues. Protective clothing must therefore be used in situations where contact would otherwise occur.

In addition to these effects, certain groups of solvents may cause problems of chronic toxicity unless exposure is controlled.

AROMATIC HYDROCARBON SOLVENTS

Toluene and xylene are commonly used solvents in rubber processes. Benzene has not itself been used in the rubber industry for many years but traces of it may be found in toluene, xylene and other hydrocarbon solvents. Benzene produces chronic effects on the bone marrow resulting in severe anaemia and leukopenia. It has also now been shown to produce cases of leukaemia. Toluene and xylene themselves do not cause the chronic blood changes associated with benzene. In the UK, the benzene content of all hydrocarbon solvents has for a number of years been voluntarily limited by the suppliers to 0.2%. In the USA, however, solvents with a greater content of benzene have been in use in recent years. Commercial toluene and xylene with less than 0.2% benzene are unlikely to lead to problems of benzene content provided that they are controlled to below their own threshold limit values (TLVs). The TLVs for the aromatic solvents are shown in Table 12.1.

SOLVENTS

TABLE 12.1
TLVs FOR AROMATIC SOLVENTS

	TLV–TWA (ppm)	TLV–STEL (ppm)	Source of data
Benzene	10	25	a
Toluene	100	150	b
Xylene	100	150	b

[a] HSE Guidance Note EH15, 1980.
[b] ACGIH, 1983.

ALIPHATIC HYDROCARBON SOLVENTS

Various aliphatic hydrocarbon mixtures are in use in the industry. These are usually defined by their boiling range, as for instance in the various 'Special Boiling Point' (SBP) solvents. At high concentrations, the major effect of this group of chemicals is narcosis, but recent work has shown that n-hexane, which is contained in varying amounts in these solvent mixtures, can cause effects on the body's peripheral nervous system (peripheral polyneuritis).[1-8] The symptoms produced may be loss of sensory perception or tremors of hands or arms. The neurotoxic effects of n-hexane are probably due to its metabolic conversion to 2,5-hexanedione[9-11] (see methyl-n-butylketone, p. 123). TLVs for individual hydrocarbons and proprietary mixtures are shown in Table 12.2.

TABLE 12.2
TLVs FOR ALIPHATIC HYDROCARBON SOLVENTS

	TLV–TWA (ppm)	TLV–STEL (ppm)	Source of data
n-Pentane	600	750	a
n-Hexane	50	—	a
Other hexane isomers	500	1 000	a
n-Heptane	400	500	a
Octane	300	375	a
Nonane	200	250	a
SBP 1	125	—	b
SBP 2	175	—	b
SBP 3	275	—	b
Petroleum rubber solvent	200	—	b
White spirit	100	—	b

[a] ACGIH, 1983.
[b] Hydrocarbon Solvent Association, 1982.

CHLORINATED SOLVENTS

Many of the chlorinated aliphatic solvents produce chronic damage to the liver and kidneys if exposure is excessive, and this may be the most important hazard in industrial use. The ability of different members of the group to cause this type of injury varies widely. Carbon tetrachloride, which has a high potential for liver and kidney damage has now for all practical purposes been abandoned by the rubber industry. Trichloroethylene, still widely used for degreasing operations and other purposes, has some potential for liver damage, but also causes central nervous system depression. Cases of cardiac arrest (ventricular fibrillation) have occurred when exposure to trichloroethylene has been followed by physical exercise. The TLVs for trichloroethylene are 50 (TWA) and 150 ppm (STEL).

1,1,1-Trichloroethane has now replaced trichloroethylene in many degreasing and solvent uses. This solvent causes little or no damage to liver and kidneys, and therefore offers definite advantages in safety over trichloroethylene. The TLVs for 1,1,1-trichloroethane are 350 (TWA) and 450 ppm (STEL).

Methylene chloride (dichloromethane) is widely used in rubber adhesives and other rubber solutions. It has a low to zero potential for liver and kidney injury, but recent work has shown that when inhaled it is converted in the body to carbon monoxide which can be demonstrated by its presence in the exhaled air and in raised levels of carboxyhaemoglobin.[12-15] Since the recognition of this effect the TLV for methylene chloride has been reduced from 500 to 100 ppm (TWA). The current TLV-STEL is 500 ppm (data from ACGIH, 1983).

Animal experiments have been carried out to determine whether members of this group of solvents possess carcinogenic potential. The International Agency for Research on Cancer has summarised the present position as follows:

Carbon tetrachloride: Probably carcinogenic in humans, sufficient evidence in animals.
Trichloroethylene: Limited evidence in animals, no data in humans.
1,1,1-Trichloroethane: Inadequate data in animals, no data in humans.
Methylene chloride: Inadequate data in animals, no data in humans.

Discounting carbon tetrachloride, it is unlikely that any of the commonly used chlorinated solvents presents a carcinogenic hazard when exposure is kept within the current TLVs.

KETONES

The ketones in common use in the industry are generally of low toxicity, with no major chronic effects. Acetone causes some eye, nose and throat irritation at concentrations above 1000 ppm, but has no significant effect below the TLV–TWA of 750 ppm (data from ACGIH, 1983). Methyl ethyl ketone (MEK) causes eye, nose and throat irritation at 350 ppm, but is not significantly irritant below the TLV–TWA of 200 ppm (ACGIH, 1982). MEK may, however, cause transient corneal damage if splashed in the eye and has caused a number of outbreaks of dermatitis in workers with excessive and repeated skin contact.

In contrast to the other ketones, methyl n-butyl ketone (MBK) has been found to cause significant chronic effects, with one well documented outbreak of peripheral neuropathy in a group of workers engaged in manufacturing printed fabrics.[16-18] 86 cases were found out of 1157 employees, the characteristic signs in those worst affected being muscle weakness in the limbs, electromyographic abnormalities, and sensory loss in hands and feet (touch, pain and temperature discrimination). Atmosphere concentrations ranged from 1 to 156 ppm, with the mean and mode below 10 ppm, but absorption by skin contact and ingestion may also have occurred. The majority of cases improved after removal of MBK from the process. The TLV–TWA for MBK (ACGIH, 1983) is 5 ppm. Animal studies have shown that the neurotoxic effects of this solvent are probably due to its metabolic conversion to 2,5-hexanedione.[9,19-21] It should be noted that this conversion can be enhanced by substances such as MEK and phenobarbitone, which increase liver microsomal enzyme activity.[22]

No reports of this type of neurotoxic effect have been made for methyl isobutyl ketone (MIBK), perhaps because the conversion to 1,5-hexanedione cannot occur. The TLV–TWA for MIBK (ACGIH, 1983) is 50 ppm, while the TLV–STEL is 75 ppm.

ALCOHOLS

Ethyl alcohol (ethanol) is a material of low general toxicity (LC_{50} (inhalation rats) 13 000 ppm). Loewy and van der Heide,[23] reporting experiments in man, observed that the inhalation of 1380 ppm caused slight symptoms of poisoning while inhalation of 5000 ppm caused heavy stupor and morbid sleepiness.[23] The current TLV–TWA is 1000 ppm (ACGIH, 1983).

Methyl alcohol (methanol) is a more hazardous solvent than ethanol, and in particular causes damage to the retina and optic nerves which may result in blindness. Chronic industrial poisoning by methanol, with marked diminution of vision, has been reported from exposure to 1200–8000 ppm for 4 years.[24] The TLV–TWA for methanol is 200 ppm, and the TLV–STEL is 250 ppm (ACGIH, 1983).

Isopropanol causes similar narcotic effects to those found with ethyl alcohol, and has been said to be about twice as toxic.[25] 400 ppm caused mild irritation of the eyes, nose and throat, and these symptoms were intensified at 800 ppm.[26]

The TLV–TWA is 400 ppm, and the TLV–STEL is 500 ppm (ACGIH, 1983).

ESTERS

In general, the common ester solvents are of relatively low toxicity although some may be irritating to the mucous membranes. The current TLVs for the simple ester solvents are shown in Table 12.3.

TABLE 12.3
TLVs FOR ESTER SOLVENTS

	TLV–TWA (*ppm*)	*TLV–STEL* (*ppm*)
Methyl acetate	200	250
Ethyl acetate	400	—
n-Propyl acetate	200	250
iso-Propyl acetate	250	310
n-Butyl acetate	150	200
sec.-Butyl acetate	200	250
tert.-Butyl acetate	200	250
Amyl acetate	100	150

GLYCOL ETHERS AND THEIR ESTERS

There are four commonly used solvents which are methyl and ethyl ethers of ethylene glycol:

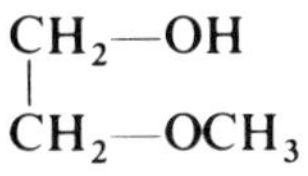

CH₂—OH
|
CH₂—OCH₃

2-Methoxyethanol, ethylene glycol monomethyl ether (EGME), methyl cellosolve, methyl oxitol

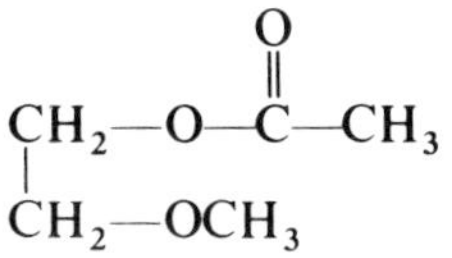

$$CH_2—O—\overset{\overset{O}{\|}}{C}—CH_3$$
|
CH₂—OCH₃

2-Methoxyethyl acetate, ethylene glycol monomethyl ether acetate (EGMEAc), methyl cellosolve acetate, methyl oxitol acetate

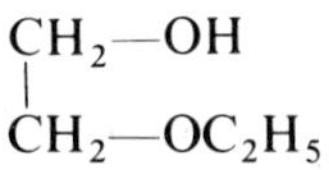

CH₂—OH
|
CH₂—OC₂H₅

2-Ethoxyethanol, ethylene glycol monoethyl ether (EGEE), cellosolve, oxitol

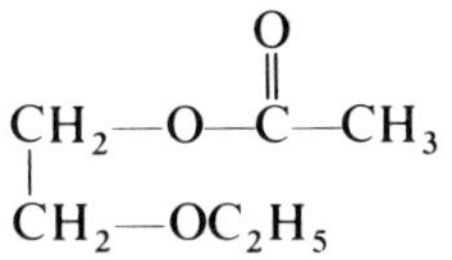

$$CH_2—O—\overset{\overset{O}{\|}}{C}—CH_3$$
|
CH₂—OC₂H₅

2-Ethoxyethyl acetate, ethylene glycol monoethyl ether acetate (EGEEAc), cellosolve acetate, oxitol acetate

The solvents are of low general toxicity (LD_{50} (oral, rat) in the range 3·4–5·1 g kg⁻¹ for all four solvents) although they may be irritants and cause central nervous system depression at high concentrations.[27] However, a number of adverse effects on animals have now been demonstrated:

Haemolytic Anaemia

An increase in red blood cell osmotic fragility has been demonstrated following exposure of rats, mice, rabbits and, to a lesser extent, guinea pigs, rhesus monkeys and humans to atmospheric concentrations of these ethers. The lowest concentrations causing this effect after 4-h exposure in rats were 2000 ppm for EGME and 62 ppm for a number of higher alkyl ethers.[28] *In vitro* experiments also demonstrated that the most potent solvent for haemolytic activity was butyl, followed by isopropyl, *n*-propyl, ethyl, with methyl the least active.[29]

Bone Marrow Suppression

Nagano and coworkers found bone marrow suppression resulting in reduced numbers of white blood cells and, to a lesser extent, red blood cells, and reduced haemoglobin concentrations after administration of the solvents to adult male mice by gavage 5 times per week for 5 weeks.[30] The lowest dose levels which gave effects are shown in Table 12.4. It can be seen from the table that in molar terms EGME and EGMEAc were about

TABLE 12.4
BONE MARROW SUPPRESSION FROM GLYCOL ETHER SOLVENTS (LOWEST DOSES FOR EFFECTS)

	Reduced numbers of white blood cells ($mg\,kg^{-1}$)	Reduced numbers of red blood cells ($mg\,kg^{-1}$)	Reduced haemoglobin level ($mg\,kg^{-1}$)
EGME	500	1 000	1 000
EGMEAc	1 000	no effect	2 000
EGEE	2 000	no effect	no effect
EGEEAc	2 000	no effect	no effect

equal in potency for this effect, and had a higher potency than the ethyl derivatives. These results were confirmed by Miller in inhalation tests on EGME at 100 and 1000 ppm.[31]

Testicular Atrophy

Nagano et al. also found dose-related decreases in testicular weight in adult male mice dosed by gavage 5 times per week for 5 weeks.[30] The lowest doses which gave effects are shown in Table 12.5. Again the methyl ether derivatives were most active. These results were confirmed by Miller in inhalation tests on EGME at 300 ppm and 1000 ppm (6 h per day for 9 out of 11 days.[31] Work by the Dow Chemical Co.[32] showed severe bilateral testicular atrophy in both rats and rabbits exposed to 300 ppm EGME for 6 h per day, 5 days per week for 13 weeks. At 100 ppm, rats showed no testicular effects, but 3 out of 5 male rabbits were reported to have moderate to severe degenerative changes. At 30 ppm, 1 out of 5 rabbits showed slight testicular changes. Infertility was reported in the male rats exposed to 300 ppm but not those exposed to 30 or 100 ppm.

TABLE 12.5
TESTICULAR EFFECTS FROM GLYCOL ETHER SOLVENTS

	Lowest dose for testicular atrophy ($mg\,kg^{-1}$)
EGME	250
EGMEAc	500
EGEE	1 000
EGEEAc	1 000

Teratological, Embryotoxic and Foetotoxic Effects

Embryotoxic and teratogenic effects from EGME and EGEE have now been demonstrated in a number of studies.[32-37] The lowest doses producing effects were:

Foetotoxicity
EGME—50 ppm (inhalation)
EGEE—400 µl kg^{-1} (oral)

Teratogenicity
EGME—50 ppm (inhalation)
EGME—31·2 mg kg^{-1} (oral)
EGEE —160-200 ppm (inhalation)

As well as these studies in animals, some effects have been noted in humans. Haematological abnormalities in workers using EGME were described by Parsons and Parsons,[38] Greenburg et al.,[39] Zavon[40] and Ohi and Wegman.[41] Psychological and neurological disorders in workers using this solvent have also been reported by Greenburg,[39] Parsons,[38] Donley,[42] Zavon,[40] Groetschel and Schürmann[43] and Ohi and Wegman.[41]

As a result of the experimental demonstration of bone marrow suppression, testicular atrophy and teratological changes at relatively low exposure concentrations, the ACGIH in 1983 reduced the TLV–TWA for all four of these solvents to 5 ppm.

The three effects listed above seem to be absent or very much weaker for higher monoalkyl ethers of ethylene glycol (propyl, butyl) and for the monoalkyl ethers of propylene glycol. A review of the toxicology of these solvents has been produced by ECETOC.[44]

REFERENCES

1. Yamada, S., *Japanese J. Ind. Health* (1967), **9**, 651.
2. Yamamura, Y., *Folia Psych. Neurol. Jap.* (1969), **23**, 45.
3. Herskowitz, A., Ishu, N. and Schaumburg, H., *New England J. Med.* (1971), 285.
4. Takeuchi, Y., Maluchi, C. and Takagi, S., *Intl. Arch. Arbeitsmed.* (1975), **35**, 185.
5. Abritti, G., et al., *Brit. J. Ind. Med.* (1976), **33**, 92.
6. Buiatti, E., et al., *Brit. J. Ind. Med.* (1978), **35**, 168.
7. Cavigneaux, A., *Securité et Hygiene du Travail* (1972), **67**, 199.
8. Gaultier, M., et al., *J. Europ. Tox.* (1973), **6**, 294.
9. DiVincenzo, G. D., Kaplar, C. J. and Dedinas, J., *Tox. Appl. Pharm.* (1976), **36**, 511.

10. Schaumburg, H. H. and Spencer, P. S., *Brain* (1976), **99**, 183.
11. Spencer, P. S. and Schaumberg, H. H., *Proc. R. Soc. Med.* (1977), **70**, 7.
12. Stewart, R. D., Fisher, T. N., Hosko, J. J., Peterson, J. E., Baretta, E. D. and Dodd, H. C., *Science* (1972), **176**, 295.
13. Stewart, R. D., et al., *Arch. Env. Health* (1972), **25**, 342.
14. Ratney, R. S., Wegman, D. H. and Elkins, H. B., *Arch. Env. Health* (1974), **28**, 223.
15. DiVincenzo, G. F., Yanna, F. J. and Astill, B. D., *Am. Ind. Hyg. Assoc. J.* (1972), **33**, 125.
16. Fontaine, R. E., Lemen, R. and Health, C. W., *PHS–CDC–Atlanta EPI* (1974), 74-39-2.
17. Billmaier, D., Yee, H. T., Allen, M., Fontaine, R. E. and O'Neill, J., *J. Occup. Med.* (1974), **16**, 665.
18. Allen, N., Mendall, J. R., Billmaier, D., Yee, H. T., Allen, M., Fontaine, R. E. and O'Neill, J. O., *Arch. Neurol.* (1975), **32**, 209.
19. Saida, K., Mendell, J. R. and Weiss, H. S., *J. Neuropath. Exp. Neurol.* (1976), **35**, 207.
20. Spencer, P. S. and Shaumberg, H. H., *J. Neuropath. Exp. Neurol.* (1977), **36**, 300.
21. Shaumberg, H. H. and Spencer, P. S., *Science* (1978), **199**, 199.
22. Couri, D., et al., *Tox. Appl. Pharm.* (1977), **41**, 285.
23. Loewy, A. and van der Heide, R., *Biochem. Zschr.* (1914), **65**, 230.
24. Henson, E. V., *J. Occup. Med.* (1960), **2**, 497.
25. Fairhill, L. T., *Industrial Toxicology*, Williams and Wilkins, Baltimore, 1949, p. 248.
26. Nelson, K. W., et al., *J. Ind. Hyg. and Tox.* (1943), **25**, 282.
27. Browning, E., *Toxicity and Metabolism of Industrial Solvents*, Elsevier, London, 1965, p. 608.
28. Carpenter, C. P., Pozzani, U. C., Weil, C. S., Nair, J. H., Keck, G. A. and Smyth, H. F., *AMA Arch. Ind. Health* (1956), **14**, 114.
29. Werner, H. W., Mitchell, J. L., Miller, J. W. and van Oettingen, W. F., *J. Ind. Hyg. Toxicol.* (1943), **25**, 157.
30. Nagano, K., Nakayama, E., Koyano, M., Dobayaski, H., Adachi, H. and Yamada, T., *Jap. J. Ind. Health* (1979), **21**, 29.
31. Miller, R. R., Ayres, J. A., Calhoun, L. H., Young, J. T. and McKenna, M. J., *Tox. Appl. Pharmacol.* (1981), **61**, 368.
32. *Release to Customers*, Dow Chemical Co., Midland, Michigan, January 1982.
33. Nagano, K., et al., *Toxicology* (1981), **20**, 335.
34. Doe, J. E., Flint, O. P., Samuels, D. H., Tinston, D. J. and Wickramaratne, A. W., *International Symposium on the Safe Use of Solvents*, Brighton, UK, March, 1982.
35. Stenger, E. G., Aeppli, L., Muller, D., Deheim, E. and Thomann, P., *Arzneim Forschung* (1971), **21**, 880.
36. Nelson, B. K., Brightwell, W. S., Setzer, J. V., Taylor, B. J. and Hornung, R. W., *Teratology* (1980), **21**, 58A.
37. Andrew, F. D., Buschbrom, R. L., Cannon, N. C., Miller, R. A., Montgomery, L. F., Phelps, D. N. and Sikov, M. R., *Battelle Report to NIOSH, Contract No. 210-79-0037*, 1981.

38. Parsons, C. E. and Parsons, M. E., *J. Ind. Hyg. and Toxicol.* (1938), **20**(2), 125.
39. Greenburg, H., Mayers, M. R., Goldwater, L. J., Burke, W. J. and Moscowitz, S., *J. Ind. Hyg. and Toxicol.* (1938), **20**(2), 134.
40. Zavon, M. R., *Am. Ind. Hyg. Assoc.* (1962), **24**, 36.
41. Ohi, G. and Wegman, D. H., *J. Occup. Med.* (1978), **20**(10), 675.
42. Donley, D. E., *J. Ind. Hyg. and Toxicol.* (1936), **18**, 571.
43. Groetschel, H. and Schürmann, D., *Arch. Tox.* (1959), **17**, 243.
44. *The Toxicology of Ethylene Glycol Monoalkyl Ethers and Its Relevance to Man*, Report No. 4, European Chemical Industry Ecology and Toxicology Centre, Brussels, July 1982.

Part III

PHYSIOLOGICAL EFFECTS OF CHEMICALS, TOXICOLOGICAL TESTING AND ATMOSPHERIC MONITORING

Part III is intended primarily as a quick reference guide to the terms and methods which may be found in Parts I and II, for the reader who may not be familiar with the nomenclature and techniques used in this field. Further details on many of the topics discussed here, each of which has been the subject of considerable study, may be found in the standard works listed in the bibliography.

13

The Effects of Chemicals on Health

Various kinds of health hazard may be caused by the chemicals used in the industry. This chapter lists the main effects, with short explanations of the terms in common use. The list of effects is not comprehensive, but covers those most commonly found when dealing with rubber chemicals. For the most part, the definitions given are intended as working guides to the terms which occur elsewhere in the book, and have been kept as short and straightforward as possible. However, the subject of carcinogenicity has been dealt with at somewhat greater length, partly because of the inherent complexity of this subject, but also in view of its importance to the consideration of rubber chemicals toxicity.

SKIN EFFECTS

Primary Irritation
Primary irritants cause a direct reddening or inflammation of the skin and mucous membranes at the site of contact. The effect may vary from a mild reversible reaction up to a serious burn, depending on many factors, including the potency of the irritant and the time of contact.

Sensitisation
When skin sensitisation takes place, the person affected becomes abnormally sensitive to a particular chemical, usually following an initial period when no problems have appeared. In some cases, this initial period can be a number of years. Once sensitised, the individual can find that even the smallest skin contact with the chemical may trigger a skin reaction, which need not necessarily be restricted to the site of contact.

EYE, NOSE OR THROAT IRRITATION

Certain of the materials used in the rubber industry may cause direct irritation of eyes, nose or throat. Irritant dust or vapours may cause irritation at all these sites, while liquid splashes in the eye may cause pain and lachrymation, and irritation or damage to the conjunctivae, iris or cornea.

RESPIRATORY CHANGES

Inhalation of vapours, dust or aerosols may cause various types of change in lung function. Some of these changes may occur shortly after a single exposure and are usually reversible, while others occur as gradual deteriorations in lung function after a long period of exposure. Such changes may be progressive even after removal from exposure as, for example, with silicosis. Some of the usual measurements taken to define lung function are the total volume of air which can be breathed out from the lungs after a full inhalation (Forced Vital Capacity, FVC), the volume of air breathed out in the first 1·0 s of such a breath (Forced Expiratory Volume – one second, $FEV_{1·0}$), the ratio between these two parameters ($FEV_{1·0}/FVC$ %) and the Peak Expiratory Flow Rate (PEFR), although other measurements are also used. The values of these parameters for an individual person depend on height, sex, age and ethnic origin. There is a gradual decline in $FEV_{1·0}$ and FVC with age, and this decline may be accelerated by exposure to certain chemicals and particularly tobacco smoke.

A few materials in use in the rubber industry, for instance phthalic anhydride and the blowing agent azodicarbonamide, are known to cause respiratory sensitisation in susceptible persons. Once sensitised, such a person will react to very small amounts of the trigger substances, far below the concentrations which would produce effects on unsensitised individuals. The effects caused are normally tightness in the chest and an asthma-like reaction, which may be immediate or delayed for several hours.

ACUTE TOXICITY

Acute toxicity has been defined as the adverse effects occurring within a short time of exposure to a chemical.[1] The term 'toxicity' is normally used here to describe the ability possessed by materials such as the systemic poisons to damage systems and organs within the body. Examples of materials causing acute toxic effects are carbon monoxide (combination of CO with haemoglobin), methyl alcohol (liver damage, damage to the retina), phosgene (lung irritation, oedema), hydrogen cyanide (interference with respiration) and ozone (irritation of mucous membranes, pulmonary oedema).

CHRONIC TOXICITY

Chronic effects occur over a longer time period than acute effects. They may be caused by repeated exposure to small amounts of a chemical, each of which is too small to cause an effect on its own. Chronic effects may be produced by heavy metals such as lead (lead poisoning), benzene (severe anaemia, leukaemia), carbon disulphide (mental disturbances, motor nerve damage, visual disturbances), aliphatic halogenated hydrocarbons such as carbon tetrachloride (damage to liver and kidneys), silica (silicosis) and certain glycol ethers (reproductive effects, anaemia).

CARCINOGENICITY

Carcinogens are those substances which cause cancer. In healthy tissues, cell reproduction is controlled to ensure that new cells are formed at just the rate required to replace old ones. This rate may be increased by certain stimuli such as hormones, or the need to replace damaged tissue (this increased proliferation is termed hyperplasia), but returns to balance when the stimulus is removed. In tumours, control of tissue growth is affected. Tumours may be benign or malignant; benign tumours may become malignant, but this is not invariably so. Benign tumours remain at the site of formation, generally retain many of the structural characteristics of the original cell, and may be within an enclosed capsule. In malignant tumours, the rate of growth is uncontrolled, and continues even when any stimulus that provoked it is removed. Malignant tumours (cancers) possess the power to invade adjacent normal tissue. The cancer cells may infiltrate blood and lymph vessels, where they may be carried to other parts of the body, be deposited, and continue to grow as secondary tumours. This process is known as metastasis. Malignant cancer cells are often markedly different in appearance from those in the parent tissue, and have partly or completely lost the specialised functions of the parent cells. Particular types of tumours have been given specific names, and some have been associated with particular causative agents, for example:

Papilloma ⎫
Adenoma ⎭ benign tumours
Sarcoma—malignant tumour of connective tissue
Angiosarcoma—malignant tumour of the liver (vinyl chloride monomer)
Mesothelioma—benign or malignant tumour of the pleura or peritoneum (asbestos)

It now seems likely that though tumours can be grouped together under the umbrella definition just given, in reality there are a number of different mechanisms at work in tumour production, and no single mode of action can be given to explain the action of all carcinogens. However, considerable evidence has now accumulated to show that the most important step in the action of many carcinogens is interference with the genetic apparatus of the cell. There are three broad stages in the carcinogenic process: metabolic activation, interaction with tissues and tumour development.

Metabolic activation is the process of modification of the carcinogen to produce a highly reactive electrophilic species (the ultimate carcinogen) which will readily react with cellular macromolecules such as DNA, RNA or protein. There may be a number of steps in this activation process. With aromatic amines and amine derivatives, the process begins with the formation of the *N*-hydroxy derivative and the complete chain has been elucidated for 2-*N*-fluoroenylacetamide:

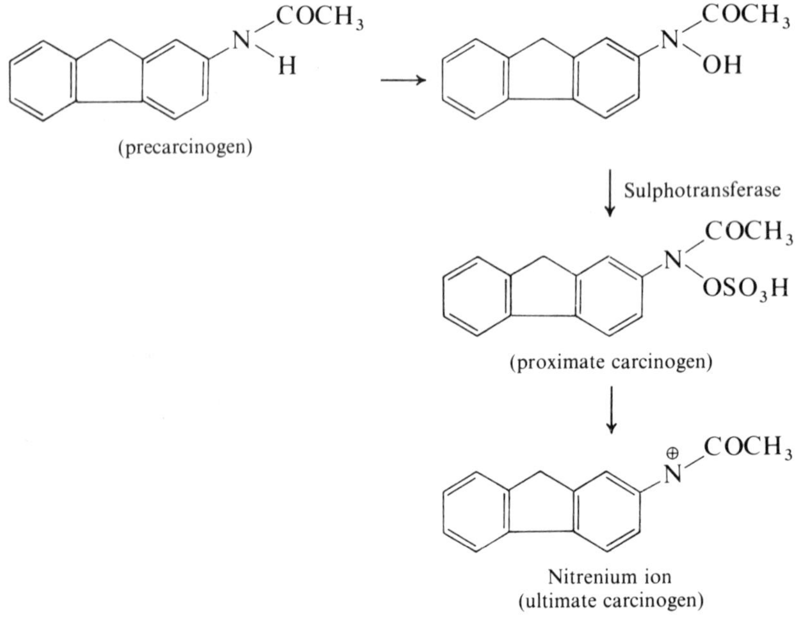

(precarcinogen) →

↓ Sulphotransferase

(proximate carcinogen)

↓

Nitrenium ion
(ultimate carcinogen)

With polycyclic aromatic hydrocarbons, a process of epoxidation is involved:

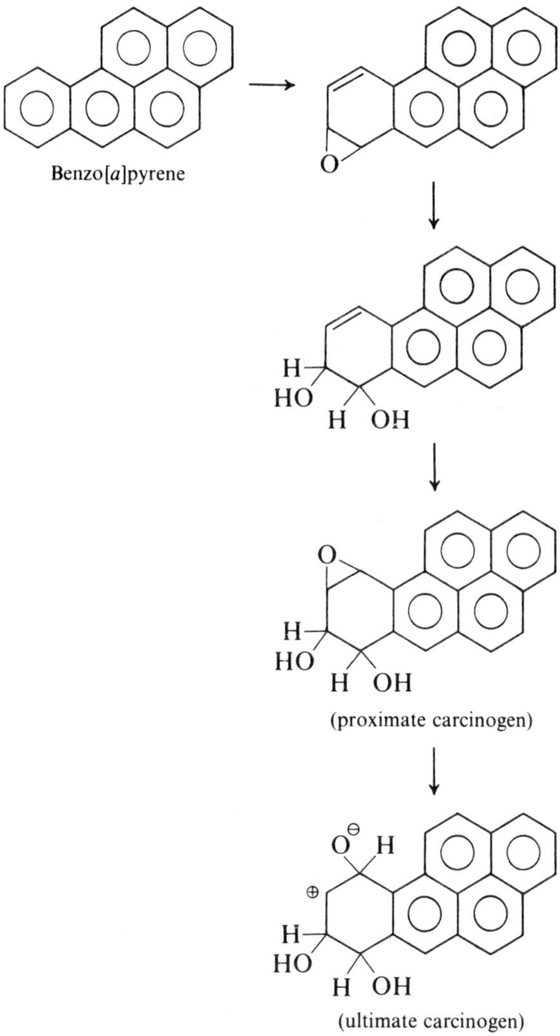

Benzo[a]pyrene
O
H
HO
H OH
O
H
HO
H OH
(proximate carcinogen)
O
H
⊕
H
HO
H OH
(ultimate carcinogen)

Most of the tissue reactions of these electrophilic groups have been studied in terms of reaction with DNA. For instance, the nitrenium ion produced by 2-*N*-fluoroenylacetamide has been shown to react at the 8-position of guanine on the DNA chain, altering the structure of the DNA molecule. The cell possesses repair mechanisms to remove such chemically induced changes, but if cell division occurs before repair is completed, a new cell with altered genetic information may result. This type of process can produce tumour progenitor cells.

The third phase in tumour development is progression of these altered cells to give a recognisable tumour. Again a number of influences are involved in this phase. Some chemicals may enhance the development of tumours from progenitor cells. For instance a single small dose of benzo[a]pyrene on the skin of mice may be insufficient on its own to cause tumour development. The area of treated skin appears normal. When this area is treated repeatedly with the non-carcinogen croton oil, even up to one year later, carcinomas and papillomas appear. Chemicals which produce this effect are called promoters. Promoters should be distinguished from co-carcinogens; these are substances which, applied at the same time as a low dose of a known carcinogen, enhance the yield of tumours. Even in the absence of promotion, development of tumour progenitor cells is affected by factors such as immunological response, hormonal status and other physiological factors and understanding of this phase is far from complete. However, whatever the mechanism, there is usually a long latent interval between exposure to a carcinogen and the appearance of a cancer. For humans, the latent interval may be of the order of 20 years.

Carcinogens which act by interfering with the genetic information in the cells have been termed 'genotoxic carcinogens', while those which act via an indirect mechanism, and do not affect the cell genetic information have been termed 'epigenetic carcinogens'.

CLASTOGENICITY AND MUTAGENICITY

The term 'clastogen' has now been used to describe the general class of substances which produce alterations in the cell chromosomes.

'Mutagens' are most commonly taken to be those substances which induce transmissable changes in the genetic material carried by male and female reproductive cells (gametes). Additionally, somatic cell mutagens are those substances which cause changes in the genetic information in other body cells; in these cases, development of the cell may be affected, although the changes will not be transmissible from parent to offspring.

It has been found that a large proportion of those materials which are mutagens are also carcinogens, and some of the recently developed short term cancer tests are in fact tests of mutagenicity potential.

TERATOGENICITY

Teratogens are substances which interfere with the normal development of the foetus during pregnancy causing malformation in the child. Probably the best known teratogen was the drug thalidomide.

OTHER REPRODUCTIVE TOXICITY EFFECTS

In addition to mutagenicity and teratogenicity, reproduction may be affected in a number of other ways by chemicals. Fertility may be reduced in males (sperm suppression, e.g. 1,2-dibromo-3-chloropropane) and in females (e.g. by lead). The foetus may experience toxic effects other than those involved in teratogenicity (foetotoxicity).

REFERENCES

1. *Principles and Methods for Evaluating the Toxicity of Chemicals. Part I Environmental Health Criteria*, World Health Organisation, Geneva, 1978.

14

Routes of Absorption of Chemicals

Chemicals can only enter the body in three ways: by skin absorption, ingestion or inhalation.

SKIN ABSORPTION

Water and most inorganic salts do not pass through the intact skin in significant quantities. On the other hand, many organic liquids are absorbed through the skin, so that this may be a significant source of exposure. Solid organic materials are usually not well absorbed through the skin, but if organic liquids such as solvents, surfactants or detergents are present, these can greatly increase the absorption of such materials.

INGESTION

Ingestion is less important than the other two routes of absorption, but may occur accidentally due to eating or smoking with contaminated hands. Some inhaled particles may also be ingested after inhalation, and this is quite an important route of exposure.

INHALATION

Inhalation can be an important route of exposure since about 10 m^3 of air is inhaled during an 8 h shift. Any contaminants present in this air can enter the body and be absorbed, though the exact mechanism by which this occurs depends on the physical form of the contaminant.

Inhalation of Dusts

The size of dust particles governs the way in which they enter the body. This can be represented schematically as follows:

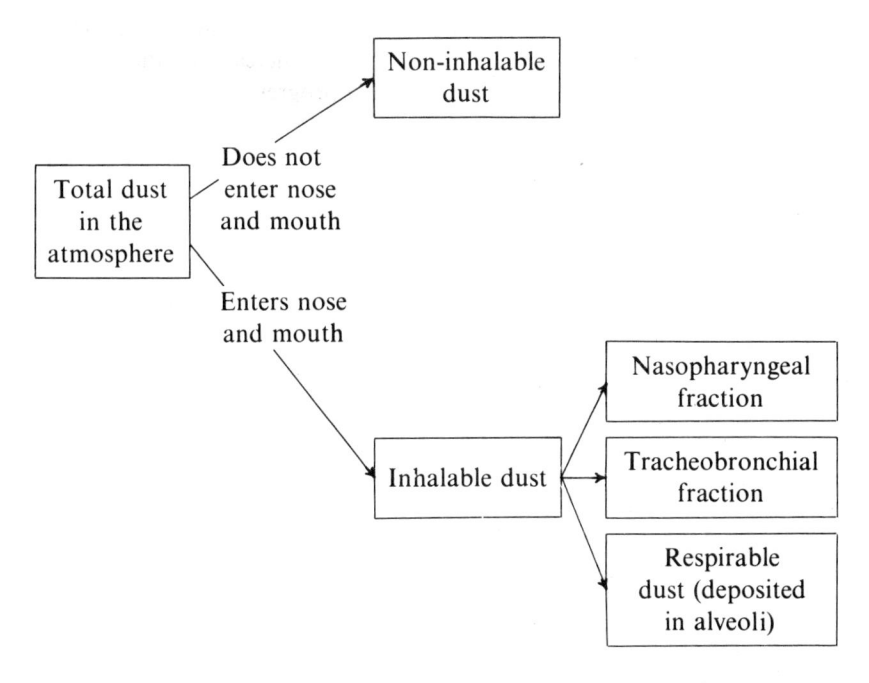

```
                           ┌──────────────┐
                    ┌─────▶│ Non-inhalable│
                    │      │     dust     │
                    │      └──────────────┘
                 Does not
┌──────────┐    enter nose
│Total dust│    and mouth
│  in the  │
│atmosphere│    Enters nose
└──────────┘    and mouth
                    │                           ┌──────────────┐
                    │                       ┌──▶│Nasopharyngeal│
                    │                       │   │   fraction   │
                    │      ┌──────────────┐ │   └──────────────┘
                    └─────▶│Inhalable dust│─┤   ┌──────────────┐
                           └──────────────┘ ├──▶│Tracheobronchial│
                                            │   │   fraction   │
                                            │   └──────────────┘
                                            │   ┌──────────────┐
                                            └──▶│  Respirable  │
                                                │dust (deposited│
                                                │  in alveoli) │
                                                └──────────────┘
```

Experiments using wind tunnels and model heads have shown that not all particles present in the air enter the nose and mouth during inhalation[1]—Fig. 14.1 gives the results of these experiments. Thus, all the small particles in the air will be inhaled, whereas only half of the particles bigger than 30 μm enter the nose and mouth. This factor of 0·5 applies up to quite large particle sizes, greater than 100 μm.

Once particles have entered the nose and mouth, they may be deposited in the nasal cavity or the region of the pharynx (*naso-pharyngeal fraction*). The larger particles in the inhaled dust will deposit preferentially in this region. Dust which penetrates past this region may deposit in the windpipe or the main airways of the lungs aided by the sticky mucus film which lines the latter surfaces (*tracheo-bronchial fraction*). All the larger particles are removed in these regions and only small particles (less than 7 μm in diameter) reach the terminal sacs (the alveoli) in the lungs (*respirable fraction*). Very small particles (less than 1 μm) tend to re-emerge from these regions with the out-going tidal air, whereas particles in the 1–7 μm range may be deposited,[5] as shown in Fig. 14.2.

Dust deposited in the alveoli can be removed only by the action of specialised blood cells (phagocytes) which are able to engulf and remove particles from these regions (phagocytosis). Certain of these dusts when inhaled in sufficient quantities may cause an increase in the fibrous connective tissues in the lung (fibrosis), with progressive deterioration in lung function.

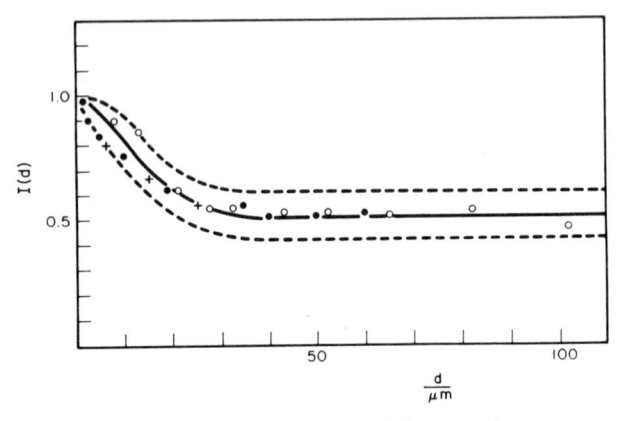

Fig. 14.1. Inhalability (I(d), fraction of particles entering nose and mouth (inhaled particles)) of particles of different sizes (d = aerodynamic diameter). ○, ref. 2; ●, ref. 3; +, ref. 4. (Reproduced from ref. 1.)

Most important of these fibrogenic dusts is silica (quartz) and minerals containing silica, but under very extreme conditions of exposure talc and carbon black can also produce fibrotic changes. Control of fibrogenic dusts is likely to involve measurements of atmospheric concentrations not only of total, but also respirable dust. Hygiene standards for respirable dust have been set for most of these materials.

Dust deposited in the higher lung passages is removed by the action of the small hairs (cilia) which line these regions and are coated with sticky mucus. The cilia constantly flick to and fro, and sweep trapped particles up to the larynx and into the oesophagus where they are swallowed. Although accumulations of dust which can cause fibrotic changes are prevented by this mechanism, dusts which are soluble in the body fluids can enter the bloodstream during clearance and may then cause systemic effects.

Inhalation of Fumes

The fumes generated by hot rubber exist as a mixture of vapours and aerosol-sized liquid particles. These particles are generally in the respirable size range and are relatively soluble in organic solvents, and presumably in body fluids. The effects of long term inhalation of this type of fume are not known for certain, but some evidence exists to suggest that this type of exposure could be associated with the production of lung cancer (see Part I, Chapter 1).

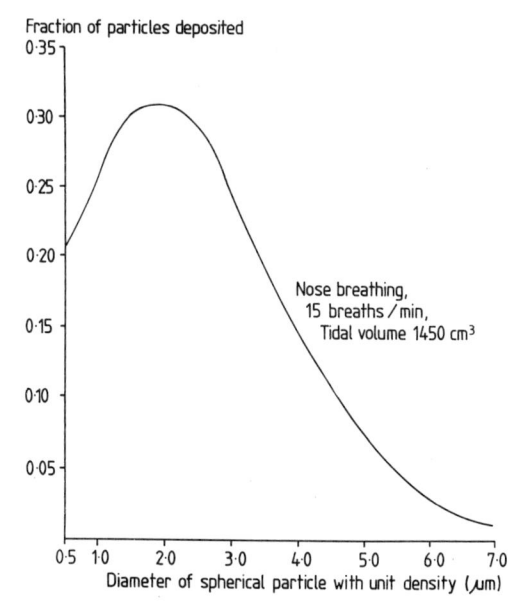

Fig. 14.2. Deposition of dust particles in the alveoli.[6,7]

Inhalation of Vapours

Many organic vapours are taken up readily in the bloodstream after inhalation, and this mode of exposure is important for these materials. The concentration of vapour in the air will be an important factor governing the extent of any toxic hazard and will be determined by the volatility of the material, the temperature of the process, the surface area of material involved and local ventilation conditions. Measurement of this concentration will usually be a first step in control of any hazards.

REFERENCES

1. Vincent, J. H. and Armbruster, L., On the quantitative definition of the inhalability of airborne dust, *Annal. Occ. Hyg.* (1981), **24**: 245–8.
2. Vincent, J. H. and Mack, D., Applications of blunt sampler theory to the definition and measurement of inhalable dust, *Inhaled Particles V*, ed. W. H. Walton, Pergamon Press, Oxford, 1981.
3. Armbruster, L. and Breuer, H., Investigations into defining inhalable dust, *Inhaled Particles V*, ed. W. H. Walton, Pergamon Press, Oxford, 1981.
4. Ogden, T. L. and Birkett, J. L., The human head as a dust sampler, *Inhaled Particles V*, ed. W. H. Walton, Pergamon Press, Oxford, 1981.
5. Orenstein, A. S. (Ed.) *Recommendations Adopted by the Pneumoconiosis Conference (Proceedings of the Pneumoconiosis Conference, Johannesburg, 1959)*, Churchill, London, 1960, pp. 619–21.
6. *International Commission on Radiological Protection Task Group on Lung Dynamics.*
7. Morrow, P. E., Bates, D. V., Fish, B. R., Hatch, T. E. and Mercer, T. T., Deposition and retention models for internal dosimetry of the human respiratory tract, *Health Physics* (1966), **12**, 173.

15

Toxicological Testing of Chemicals

Although it is impossible to be sure that all the potential human effects of a new chemical have been discovered before it is put into use, a good deal of information on the likely hazards can be obtained by tests on animals or simpler life forms. Many of the potential health hazards to humans can be recognised and prevented by such tests. Higher mammals, for instance dogs or monkeys, are likely to be more representative of human response than are tests on lower organisms, since the metabolic pathways in these mammals are usually nearer to those of the human system. However, where such tests have to be continued for a large proportion of the animal's life, it may take a number of years to complete an experiment. Such tests will not provide quick answers and are inevitably expensive. For this reason many tests are now made with short-lived mammals such as rats or mice, and for mutagenicity (carcinogenicity) tests a number of systems using bacterial or other cultured cells have been developed. Whatever animal or cell system is used to provide toxicological data, there are numerous problems and pitfalls in extrapolating the results to man, and expert advice will be required before any conclusions are drawn regarding the implications of these tests.

DOSE/RESPONSE RELATIONSHIPS

Measurement of the toxicological properties of chemicals is inherently more difficult than measurement of a simple physical property. In any group of animals or numans, there will be a variation in susceptibilities to the effects of a chemical. Most of the individuals in the group will respond similarly to a central dose level, but a few will respond to much lower levels, while a few will only respond at much higher levels. If the percentage of the

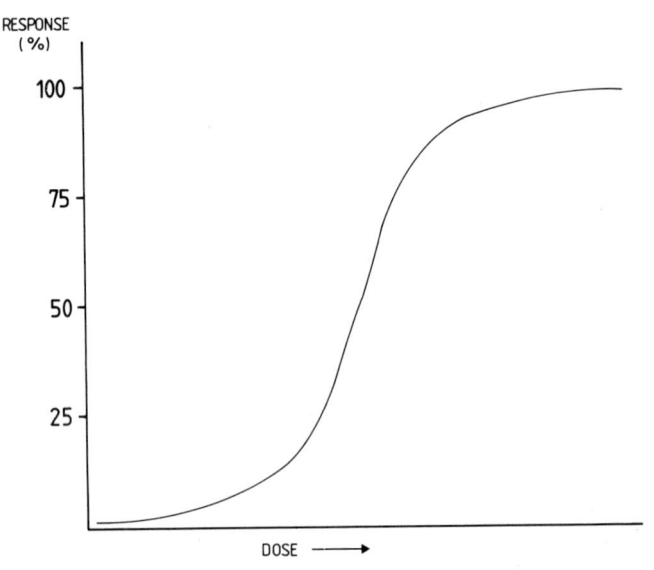

Fig. 15.1. Dose/response curve from animal tests.

group who respond to a given dose is plotted as a cumulative frequency against dose, a sigmoidal or S-shaped curve commonly results, as shown in Fig. 15.1. The central portion of such a curve approximates to a linear response; the minimal portion is commonly asymptotic to 0% and the maximal portion to 100%.

Such S-shaped frequency curves can be transformed to give straight lines by plotting the % response data on a scale of 'probability units', as in Fig. 15.2. This treatment makes such data easier to evaluate. As can be seen from the figure, the slopes of such dose/response curves can be different for different chemicals, and this is of considerable importance when evaluating the data obtained from animal tests. For instance, the two chemicals shown here produce equal effects at the point where the lines cross, and at this dose 50% of a test group would be affected by either chemical. If these results were obtained from an LD_{50} experiment, the chemicals would be shown to have equal LD_{50} values. However, at higher doses chemical A would affect more animals than chemical B, while at lower doses the reverse would be true. The slopes of such dose/response curves are therefore of great importance in relation to margins of safety, etc.

Many of the test routines described in this chapter are taken from the standard tests now recommended by the European Economic Community. Details of these tests can be found in refs 1 and 2.

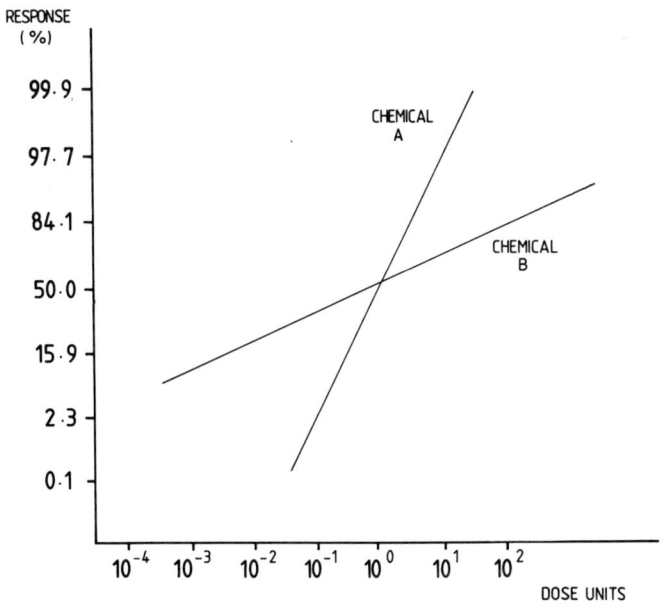

Fig. 15.2. Dose/response curves plotted on log/probability scales.

ACUTE TOXICITY TESTING

Acute toxicity has been defined as the adverse effects occurring within a short time after administration of a single dose or multiple doses given within 24 h.

LD$_{50}$

This is determined by administering a range of doses to animals in order to find the dose which will kill half of the experimental group. The animals normally used are mice, rats, guinea pigs or rabbits. One group is used for each dose level; it will often consist of 5 males and 5 females of one species. The chemical may be given by mouth or through a stomach tube (oral administration), by subcutaneous or intraperitoneal injection, by skin application or by inhalation. The single dose which will kill half of an experimental group within, say, 14 days of administration, is known as the 50% lethal dose (LD$_{50}$) and is usually given in mg kg^{-1} of body weight. For inhalation dosing, however, this dose is usually known as the 50% lethal concentration (LC$_{50}$) and is usually given as a simple atmospheric concentration in mg m^{-3}. Since there is a considerable variation of results from such testing, depending on the animal used, the administration route and the exact procedures used in the test, it is best if some indication of these variables is given when quoting a figure—e.g. LD_{50} (oral, rat) = 510 mg kg^{-1}, LC_{50} (4 h, rat) = 250 mg m^{-3}, LC_{50} (1 h, mouse) = 75 mg m^{-3}. It is also now common practice to give 95% confidence limits to any figure quoted. Materials have been classified into various grades within the EEC using the oral LD_{50} figure.

Oral LD_{50}	Rating
< 25 mg kg^{-1}	Highly toxic
25–200 mg kg^{-1}	Toxic
200–2000 mg kg^{-1}	Harmful

Although LD_{50} figures are commonly available, the test is essentially a range-finding procedure for determining acute toxic hazard, and more sophisticated testing will be required to assess the potential problems of the chemical under more realistic conditions of exposure.

Skin and Eye Irritation Tests

Primary Skin Irritation

Primary skin irritation potential is assessed by applying the chemical to the clipped skin of animals, usually rabbits, for 4 h. The chemical may either be used neat, or be diluted in water or another suitable vehicle. The application is normally covered with surgical gauze and semi-occlusive bandage during the exposure period to reduce evaporation and prevent removal. After the exposure period, the bandages are removed and the skin washed to remove any residual test substance. The skin is then observed at intervals, and the severity of any irritation noted after the first hour and then every 24 h. Various scales have been used to record these subjective assessments, the one shown in Table 15.1 now being in use within the EEC (p. 13, ref. 1).

Skin Sensitisation

Skin sensitisation is usually assessed by a two-part test—induction and challenge. For instance, in the Magnusson and Kligman maximisation test,[3] an irritant concentration of the test substance is injected intradermally into the shoulder area of guinea pigs, and one week later the test substance is applied to the skin at the injection site (induction). The exposure site is covered for 48 h. A non-irritant concentration of the test substance is applied to the treated area of the animals two weeks after the

TABLE 15.1
EFFECTS OF CHEMICALS ON THE SKIN

Observed reaction	Score
Erythema and eschar formation	
No erythema	0
Very slight erythema (barely perceptible)	1
Well defined erythema	2
Moderate to severe erythema	3
Severe erythema (beet redness) to slight eschar formation (injuries in depth)	4
Oedema formation	
No oedema	0
Very slight oedema (barely perceptible)	1
Slight oedema (edges of area well defined by definite raising)	2
Moderate oedema (raised approximately 1 mm)	3
Severe oedema (raised more than 1 mm and extending beyond the area of exposure)	4

induction treatment (challenge) and the skin reaction is subjectively assessed at 24, 28 and 72 h using a similar scale to that given for primary irritation.

Eye Irritation
Eye irritation is assessed by applying a small quantity of the chemical to one eye of each of the test animals, usually rabbits. Half the animal group have the chemical washed out after 1 min. The eyes are then observed for up to seven days for evidence of irritation and any inflammation rated on a subjective scale (p. 15 of ref. 1). Assessments are made separately for cornea, iris, conjunctivae and lids and/or nicitating membrane.

The test is not applicable to strongly acid (pH 2) or alkaline (pH 11·5) substances, nor to severe skin irritants.

The effects of remedial irrigation may be assessed by washing the treated eye with tap water at different time intervals after installation of the test chemical, and observing the effects.

SUB-ACUTE AND SUB-CHRONIC TOXICITY

Sub-acute and sub-chronic toxicity are normally assessed by extended feeding, skin application or inhalation tests. Commonly used methods are the 28- and 90-day feeding trials. In these tests, groups of animals are fed known quantities of the test chemical in their diet continually for the period of the test. During this time they are weighed and any observable effects noted. Haematology and blood chemistry (liver and kidney function) are usually carried out at the end of the dosing period. The animals are then sacrificed, and a necropsy carried out to find whether any changes in the body organs have taken place. Histology (microscopic examination of the detailed cellular structure) of the organs and tissues is carried out to further identify changes.

This type of test will usually give more detailed information on the physiological effects of exposure to a chemical than will the LD_{50} test. It may also provide information on the dose required to cause specific changes. Doses must be selected so that the dose that produces no observed effects in the animal has been established.

CARCINOGENICITY TESTS

Conventional carcinogenicity tests are carried out by administering high doses of the test chemical to a group of animals over a long period, usually a significant proportion of the animal's lifetime. In the highest dosage group, the dose used will commonly be close to the maximum that the animal can tolerate (the MTD).

Any animals that die during the test are examined for organ changes and at the end of the test the animals are sacrificed and similar examinations made. Any tumours found are recorded. A control group of animals is used to provide information on the background rate of tumours in the species used for the test.

It is impossible to set down fixed rules for interpretation of the results from these tests, since so many variables contribute to each set of results. Instead, carcinogenicity bioassay results need to be carefully assessed by many experts in different fields on a case-by-case basis. Such a procedure is used by the International Agency for Research on Cancer. The IARC has adopted a classification system which places chemicals and processes into four groups:

(i) sufficient evidence;
(ii) limited evidence;
(iii) inadequate evidence;
(iv) no data.

This assessment is used, together with epidemiological data, to evaluate the carcinogenic risk to humans. The IARC has defined three groups of chemicals and processes on this basis:

Group 1: The chemical, group of chemicals, industrial process or occupational exposure is carcinogenic to humans.

Group 2: The chemical, group of chemicals, industrial process or occupational exposure is probably carcinogenic to humans.

Group 3: The chemical, group of chemicals, industrial process or occupational exposure cannot be classified as to its carcinogenicity in humans.

At the present time, the IARC lists for Groups 1 and 2 are as shown in Table 15.2.[4]

As has previously been stated, conventional carcinogenicity bioassays can take many years (7–9 years if dogs are used) and are very expensive. Various short term carcinogenicity tests have been developed to speed up this type of testing.

Short Term Carcinogenicity Tests

It has been found that those chemicals which are capable of causing genetic damage, usually recognised as mutagenic changes in the organism, are also likely to possess carcinogenic activity. The majority of the short term cancer tests which have been developed over the last few years are in fact tests for mutagenic activity. They are of two main types: the first group detects point mutations in genes, usually in bacterial cultures; the second group looks for recognisable visual damage to chromosomes in mammalian cells either *in vivo* or *in vitro*. Short term tests of this kind have proliferated in recent years, each with their own degree of success as judged by how well the results compare with those obtained by conventional animal tests. Four of the better known tests will be described in more detail.

The Ames Test

This test looks for mutations caused by the test chemical in specially developed mutant strains of *Salmonella typhimurium* bacteria. These strains contain a defect which prevents the bacteria from synthesising the amino acid histidine, which is required for bacterial growth. The strains will therefore only grow if histidine is supplied to them. The defect in these bacteria can be reversed by chemicals which have mutagenic activity and the bacteria will then grow normally even in a histidine deficient culture. (Because of this, the test is sometimes described as a 'reverse mutation

TABLE 15.2
IARC ASSESSMENTS FOR CARCINOGENICITY

Group 1: The Working Group concluded that the following seven industrial processes and occupational exposures and 23 chemicals and groups of chemicals are causally associated with cancer in humans.

Industrial processes and occupational exposures:
  Auramine manufacture
  Boot and shoe manufacture and repair
    (certain occupations)
  Furniture manufacture
  Isopropyl alcohol manufacture
    (strong-acid process)
  Nickel refining
  Rubber industry (certain occupations)
  Underground haematite mining
    (with exposure to radon)

Chemicals and groups of chemicals:
  4-Aminobiphenyl
  Analgesic mixtures containing phenacetin[a]
  Arsenic and arsenic compounds[a]
  Asbestos
  Azathioprine
  Benzene
  Benzidine
  N,N-Bis(2-chloroethyl)-2-naphthylamine (Chlornaphazine)
  Bis(chloromethyl)ether and technical-grade chloromethyl methyl ether
  1,4-Butanediol dimethanesulphonate (Myleran)
  Certain combined chemotherapy for lymphomas[a] (including MOPH[b])
  Chlorambucil
  Chromium and certain chromium compounds[a]
  Conjugated oestrogens[a]
  Cyclophosphamide
  Diethylstilboestrol
  Melphalan
  Methoxsalen with ultra-violet A therapy (PUVA)
  Mustard gas
  2-Naphthylamine
  Soots, tars and oils[a,c]
  Treosulphan
  Vinyl chloride

Group 2: The following 61 chemicals, groups of chemicals or industrial processes are *probably* carcinogenic to humans

Group 2A:
  Acrylonitrile
  Aflatoxins
  Benzo[a]pyrene
  Manufacture of magenta[a]
  Nickel and certain nickel compounds
  Nitrogen mustard

Beryllium and beryllium compounds[a]
Combined oral contraceptives[a]
Diethyl sulphate
Dimethyl sulphate

Group 2B:
Actinomycin D
Adriamycin
Amitrole
Auramine (technical grade)
Benzotrichloride
Bischloroethyl nitrosourea (BCNU)
Cadmium and cadmium compounds
Carbon tetrachloride
Chloramphenicol
1-(2-Chloroethyl)-3-cyclohexyl-1-nitrosourea (CCNU)
Chloroform
Chlorophenols (occupational exposure to)[a]
Cisplatin
Dacarbazine
DDT
3,3'-Dichlorobenzidine
Dienoestrol
3,3'-Dimethoxybenzidine (*ortho*-Dianisidine)
Dimethylcarbamoyl chloride
1,4-Dioxane
Direct Black 38 (technical grade)
Direct Blue 6 (technical grade)
Direct Brown 95 (technical grade)

Oxymetholone
Phenacetin
Procarbazine
ortho-Toluidine

Epichlorohydrin
Ethinyloestradiol
Ethylene dibromide
Ethylene oxide
Ethylene thiourea
Formaldehyde (gas)
Hydrazine
Mestranol
Metronidazole
Norethisterone
Oestradiol-17β
Oestrone
Phenazopyridine
Phenytoin
Phenoxyacetic acid herbicides (occupational exposure to)[a]
Polychlorinated biphenyls
Progesterone
Propylthiouracil
Sequential oral contraceptives[a]
Tetrachlorodibenzo-*para*-dioxin (TCDD)
2,4,6-Trichlorophenol
Tris(aziridinyl)-*para*-benzoquinone (Triaziquone)
Tris(1-aziridinyl)phosphine sulphide (Thiotepa)
Uracil mustard

---

[a] The compound responsible for this carcinogenic effect in humans cannot be specified.
[b] Procarbazine, nitrogen mustard, vincristine and prednisone.
[c] Mineral oils may vary in composition, particularly in relation to their content of carcinogenic polycyclic aromatic hydrocarbons.

assay'.) The test therefore consists of exposing large numbers of these bacteria to a known quantity of the test chemical, incubating at 37 °C for 72 h and counting the number of revertant colonies which have grown.

A positive result is one where a statistically significant excess number of colonies is produced compared with the number given by controls. The bacterial strains used by Ames also contain a mutation which makes the cell walls more permeable to the test chemical, and do not contain the DNA repair mechanisms by which normal cells protect themselves from the major effects of mutagens.

Many chemicals are only mutagenic (or carcinogenic) after they have been chemically altered by enzyme systems within the mammalian body. To allow (as far as possible) for this effect, a preparation rich in these enzymes is added to the bacterial culture in the Ames test. This preparation (supplemented liver fraction, S-9) is made from the livers of rats which have been treated with a compound known to induce a high level of enzyme activity.

The test is normally carried out on a range of concentrations of the test chemical, together with untreated controls and positive controls treated with a known mutagen.

A similar reverse mutation assay has been developed using tryptophan reversion in special strains of *Escherichia coli* bacteria.

The Micronucleus Test
This is a test of chromosome damage in the bone marrow of mice or rats exposed to the test compound. When chromosome damage has been caused in a cell, fragments of the chromosomes may not be included in the cell nucleus at the next cell division. These fragments remain in the cells as single or multiple micronuclei. They are easily seen in red blood cells (erythrocytes), and these are used in the test. To examine the effects of the test chemical, only young cells, less than 24 h old are examined. These cells can be recognised by the fact that they stain blue with Giemsa, due to the presence of RNA in the cells, whereas older cells stain pink. Blue staining red blood cells are termed 'polychromatic erythrocytes' and the test looks for micronucleated cells of this type.

In practice, the test chemical is given orally in a range of doses up to the maximum tolerated dose. The animals are killed at fixed periods after dosing is complete, and bone marrow smears are taken. The slides are stained with Giemsa, and the number of micronucleated cells per 1000 polychromatic erythrocytes counted under a light microscope. If the count for treated animals is twice that for controls, a positive response is claimed.

Chromosome Damage (Clastogenicity) in Cultured Mammalian Cells (in vitro *Mammalian Cytogenetic Test*)
The damage caused to chromosomes by mutagens can often be recognised under the microscope as physical changes such as breaks and other aberrations. These changes are most easily seen when the cell is about to divide and the chromosomes are contracted and positioned at the centre of the cell (metaphase stage). The test uses cultures of Chinese hamster cells or cultures of human lymphocytes taken from a blood sample. The cells are incubated in a culture medium at 37 °C and the test chemical is then added, either with or without supplemented liver fraction (S-9 mix). A further period of incubation takes place, and a spindle inhibitor such as the alkaloid colchicine is added to accumulate cells at the metaphase stage. The cells are then centrifuged, fixed and stained. They are then examined under the microscope for chromosomal abnormalities, in comparison with positive and negative control cultures. For a positive response it should be possible to demonstrate a statistically significant dose related increase in the number of aberrations.

In vivo *Bone Marrow Cytogenic Test*
This test looks for chromosome damage in bone marrow cells in animals which have been treated with the test chemical. Rodent species such as rats, mice or Chinese hamsters are used. They are treated with the test chemical, usually once only. After a suitable interval, centred on 24 h, the animals are injected intraperitoneally with a spindle inhibitor such as colchicine to obtain an adequate number of cells in the metaphase state. They are then sacrificed and bone marrow smears are taken. At least 50 metaphases are then examined under the microscope for aberrations such as gaps, breaks and interchanges. The data obtained are compared with those from control groups.

REFERENCES

1. *Notification of New Substances Regulations, 1982: Approved Code of Practice No. 10, Methods For The Determination of Toxicity*, HMSO, London, 1982.
2. *Directive 79/831/EEC, Annex V*, European Economic Community, 18 September 1979.
3. Magnusson, B. and Kligman, A. M., *Allergic Contact Dermatitis in the Guinea Pig: Identification of Contact Allergens*, Charles C. Thomas, Springfield, Illinois, 1970.
4. *IARC Monographs, Supplement 4*, International Agency for Research on Cancer, Lyon, France, 1982.

16

Atmospheric Monitoring Methods

For most of the materials used in the rubber industry, inhalation of dusts, fumes and vapours is the most important route of exposure. The amount of these materials present in the atmosphere must therefore be measured to find whether a hazard exists. Various limits are used to tell whether the result indicates a hazardous situation. In the United States, the American Conference of Governmental Industrial Hygienists (ACGIH) has for a considerable number of years produced an annual list of *Threshold Limit Values* (TLVs) and these have also been extensively used by the Health and Safety Executive (HSE) in the UK. A few materials have been assigned specific UK *Control Limits* by the Advisory Committee on Toxic Substances (ACTS), which has representatives from the HSE, industry and trades unions. A more comprehensive list of specific UK *Exposure Limits* is now being proposed by HSE and ACTS. A number of other countries have produced national lists of limits, notably the West German *MAK* list.

The most used of these limits has been the ACGIH's TLV list, which has been reproduced each year by the HSE as Guidance Note 15.[1]

THRESHOLD LIMIT VALUES

Two types of limit are used by the ACGIH. The *time-weighted average TLV* (TLV–TWA) defines the maximum average exposure concentration over the 8-h shift. For gases or vapours it is usually given in units of parts per million by volume (ppm)—i.e. 100 ppm is 100 volume units of gas or vapour per million volume units of air. For dusts and aerosols (small solid or liquid particles dispersed in air) it is usually given in units of milligrams of dust or aerosol per cubic metre of air (mg m^{-3}). Although the TLV–TWA defines the maximum average concentration over the 8-h shift, it is usually permissible to exceed this value for short periods of time provided that these 'excursions' are balanced out by periods below the TWA value. However, for some materials, which have severe irritant properties or are acutely toxic at concentrations close to the TLV–TWA a *Ceiling* notation is added to the TLV–TWA. Materials with this ceiling notation must be held below the ceiling value (which is numerically the same as the TLV–TWA value) at all times, however brief the time period may be.

For materials that do not have a ceiling value, limits are placed on any excursions above the TLV–TWA by the *short-term exposure limit TLV* (TLV–STEL). The STEL value defines the upper limit of concentration which must not be exceeded in any time period, however short. Each excursion into the region between the TWA and the STEL must last for less than 15 min, and a maximum of 4 of these excursions is allowed per day, with at least 1 h between excursions.

The evidence used when setting the TLVs is set out by the ACGIH in the *Documentation to the Threshold Limit Values*.[2] The ACGIH point out that the TLVs define concentrations to which it is believed that nearly all workers may be repeatedly exposed day after day without adverse effect, but that a small percentage of workers with greater individual susceptibility may experience discomfort at or below the TLV, and an even smaller percentage may be affected more seriously by aggravation of an existing condition or by development of an occupational illness. For this reason it can be seen that TLVs do not define sharp divisions between safe and dangerous concentrations, and that wherever possible exposure concentrations should be reduced to the lowest practicable level.

Only a handful of the several hundred rubber chemicals have so far been assigned official TLVs. The BRMA has recommended limits for many of the common solid rubber chemicals in its code of practice on *Toxicity and Safe Handling of Rubber Chemicals*.[3] The chemicals are divided into two groups, the less hazardous group being given a limit of $10\,\mathrm{mg\,m^{-3}}$ in line with nuisance dust, while the more hazardous group are given a limit of $2{\cdot}5\,\mathrm{mg\,m^{-3}}$.

STATIC AND PERSONAL SAMPLING

Static (or area) samples are obtained at a fixed position and give the concentration over the measuring period at this position. Although samples of this type can be useful for producing a contamination map of a work area when designing or checking control equipment, they do not usually give a measure of exposure for any worker in this area, since most rubber industry jobs involve considerable movement around the work station.

In order to get a true measure of an individual worker's exposure to an atmospheric contaminant it is best to use miniature sampling equipment which can be carried by the worker during his normal operations. The sampling head of the equipment should be placed as close as practicable to his breathing zone, normally on his collar or lapel. This type of sampling is termed personal sampling. Almost all the equipment available for personal sampling is designed to give a time-weighted average result, so that sampling should be carried out over a reasonable fraction of the working shift. A few personal samplers now produce a continuous record of exposure over the time period in the form of a concentration versus time graph. These can be used to check whether the ceiling value or STEL has been exceeded at any time. For many materials, however, no equipment is available to give personal results in this form, and estimations must be made either from the results of continuous static monitors, or by judging when high personal exposure is likely to occur, so that sampling can be carried out over this period.

The hygienist taking these measurements should wherever possible supervise the sampling operation over the whole of the measurement period, so that the exact details of each sample can be recorded. This supervision makes it possible to identify abnormal results caused by such circumstances as burst bags of raw materials and accidental or even deliberate interference with the sampling equipment, which may otherwise cause false conclusions to be drawn. It is also important to take sufficient samples to enable a reliable conclusion to be reached, especially where the results are used to design control systems involving considerable capital expenditure.

DUST MEASUREMENT

Total Dust

The classic method for measuring dust concentrations is to draw a known quantity of air through a weighed filter and then reweigh it. For simple time-weighted average measurements this type of monitoring has now taken over from the earlier particle counting methods for all materials except asbestos and possibly man-made mineral fibres. For such time-weighted average measurements sampling should be carried out for a reasonable proportion of the normal shift.

Static samples of this kind may be taken with a mains operated sampling pump, and flow rates in the range 15–25 litres min^{-1} are typically used for such dust measurements. For personal samples, a small pump powered by rechargeable Ni/Cd batteries is used, and flow rates of 1–2 litres min^{-1} are in common use.

The sampling pump must be capable of giving a steady flow against the resistance of the filter for periods of up to 8–10 h. Many modern personal samplers incorporate a sensor which detects the rate of air flow and automatically adjusts the motor speed to compensate for changes in the resistance of the filter as the sample is collected—these samplers are known as stabilised flow samplers. Even with these samplers flow rates should be measured at the inlet to the filter before sampling begins and at intervals during the sampling period.

The holder used for the filter may affect the result obtained. Ideally a holder should be unaffected by its orientation, since this is likely to alter as the wearer moves, and should protect the filter against accidental interference. The 'modified UKAEA holder'* goes some way towards these ideals.

Filter holders which admit only 'inhalable' dust (see Chapter 14, p. 141) are now being designed.

Respirable Dust

The same technique is used for respirable as for total dust, except that the dust is passed through a separator of some kind which removes the larger particles and only passes through to the filter those particles which are small enough to penetrate to the lower lung. The separator must produce the same type of sorting action on the range of particle sizes that the lung itself does (see Chapter 14, pp. 141–2 and two main types are in use, the elutriator and the cyclone. The elutriator consists of a set of parallel horizontal plates over which the dust-laden air stream is passed. In this laminar air flow the larger particles settle onto the plates while the respirable particles are passed through to the filter. Elutriators are commonly used on mains and battery operated static samplers. Personal samplers for respirable dust use miniature cyclones to separate the larger dust particles. In the commonest form of these the upwards moving air stream is made to swirl and large dust particles fall into the centre of this 'whirlpool' and are deposited in a rubber grit pot. Respirable particles continue with the upwards moving air stream to meet the bottom face of the

* Sources of equipment mentioned in this chapter are listed in the Appendix, p. 176.

filter. Both elutriators and cyclones must be operated at precisely their designed flow rates to obtain the correct sorting characteristics. The air flow must not be pulsating and where diaphragm or piston pumps are used a flow damper must be incorporated in the flow line.

Some estimate of total dust can be made by weighing the grit pot before and after sampling, but such measurements are likely to be very inaccurate because of the relatively large tare weight of the grit pot (compared to the small dust weight) and because of hygroscopic changes in the weight of the rubber. Samples can sometimes be washed from the grit pot, but this is again difficult with many of the materials in use in the rubber industry. In general, it is recommended that the miniature cyclone is used only to measure respirable dust and separate, perhaps concurrent, samples are taken for total dust as previously described.

Filters

A variety of filter types is available for use in dust sampling work and the type to be chosen for a given job will depend both on the requirements of the sampling conditions and on the subsequent analytical treatment (if any) which is to be used. For simple gravimetric work it is important that the filter is not greatly affected by changes in the atmospheric humidity, and for this reason paper filters cannot be used, and some types of cellulose esters are not recommended. The main filter materials for this type of work are glass fibre, silver, regenerated cellulose, cellulose nitrate, PVC, nylon, teflon and polystyrene.

Some types of synthetic filter (e.g. PVC and some cellulose esters) develop an electrostatic charge during sampling, and steps must be taken to eliminate this charge before weighing. If this is not done the weighing process may be seriously affected. Static eliminators can be obtained to remove the charge, or alternatively the filters may be pretreated with materials to prevent charge accumulation.

Use of silver filters in rubber factories is not recommended because of reaction with the sulphides generated during the curing process.

Weighing Equipment

The total weight of dust collected by personal sampling is often very small, perhaps as little as 100 μg. A microbalance capable of weighing down to at least 10 μg is required to accurately weigh these quantities of dust. Although it is possible to use a traditional chemical microbalance for this work, this type of balance needs carefully controlled surroundings and an experienced operator to produce good results. It also tends to be slow in use. The newer types of electronic microbalance have therefore been widely adopted for dust monitoring measurements. These balances are more tolerant of vibration and temperature changes and are quicker in use.

Automated Dust Monitoring

All the previously described equipment produces a single average dust concentration measurement over a period of several hours. Instruments are now being produced which will give a much quicker reading of the dust concentration present at an area without the need for weighing operations. Three main types, all using different physical principles, are now available.

β-Radiation Attenuation

Various versions of these portable battery operated samplers are available to give continuous respirable or average dust concentrations. In the commonest version, the airstream is taken through a cyclone, to remove particles greater than 10 μm diameter, and then through a circular nozzle which causes the airstream, and any particles present in it, to impact on to a coated Mylar disc. A beam of β-radiation is passed through the collected dust sample on the disc and the proportion of radiation stopped by the dust is a direct measure of the mass of the dust which has been deposited. The dust concentration is displayed automatically on a three-digit LED display in units of mg m^{-3}. The disc can then be turned to a fresh position to enable a subsequent sample to be taken, with up to 95 samples per disc being possible.

Two versions of the instrument are available, with automatic sampling times of 1 and 4 min respectively, and a range switch to cut the sampling time by a factor of 10. The instrument can also be operated semi-manually for longer time periods. The measuring range for the 1-min instrument is 1–150 mg m^{-3} and for the 4-min instrument 0·2–50 mg m^{-3}. Both models will measure concentrations down to 0·02 mg m^{-3} when used in the semi-manual mode for 10-min samples. A more sophisticated version of this instrument, in which the disc is automatically moved to give continuous measurements, is also available.

In all these instruments the system of dust impaction on the Mylar disc is inefficient for particles smaller than 0·9 μm and they therefore give a poor response to fumes or clouds of very fine particles. The system is, however, efficient for the particle sizes of most physiological importance. The instruments are sold as GCA–RDM automatic dust monitors.

Vibrating Quartz Crystal

This is a battery operated sampler which measures respirable dust concentrations by electrostatic deposition on a vibrating quartz crystal. The rate of vibration of the crystal is slowed down in direct proportion to the weight of dust deposited, and this slowing is measured by comparison with a reference crystal. An impactor can be used to remove large particles (>3.5 μm) from the air stream before passing to the quartz crystal and the readings then approximate to respirable dust.

The readings are automatically displayed in units of mg m^{-3}. Sample times of 24 or 120 s can be used, giving a measurement range of 0·005–10 mg m^{-3}. For a given sampling period therefore, the instrument is more sensitive than the β-radiation instruments previously described. After a number of readings have been taken the quartz crystal must be cleaned by turning a dial which moves detergent-coated sponges over the crystal.

The electrostatic collection system used in this instrument is efficient for small particles, down to 0·01 μm, and the instrument will therefore give readings in fume or fine dust clouds. The impactor used to remove large particles has sharper cut-off characteristics than the experimentally-determined human airways/lung retention curve, and may for this reason introduce some inaccuracy. Even when using no impactor the instrument may begin to be inefficient for large particles and may not give agreement with total dust figures—it is in any case not really intended for this type of measurement. This monitor is sold as the TSI-3500 automatic dust monitor.

Narrow Angle Forward Light Scattering

Two monitors are now available. The first (Simslin) has been developed by the Safety in Mines Research Establishment in the UK, now part of the Health and Safety Executive. It uses an elutriator to separate respirable particles and these are then passed through the optical cell of the instrument where they cause scattering of a laser light beam. The portion of the light scattered at a narrow forward angle is collected and measured and this signal serves to provide the read-out for dust concentration, which is displayed digitally at 15 s intervals in terms of mg m^{-3}. A running average dust concentration over a 15 min period can also be displayed. The output from the instrument is stored on interchangeable solid state memory units, or can be fed immediately to a recorder. After passing through the optical cell, the dust laden airstream is passed through a filter, and the weight of dust collected on this filter at the end of the sampling period can be used to

TABLE 16.1
FEATURES OF AUTOMATIC SAMPLERS

Advantages	Disadvantages
Speed	Too large to use as personal samplers
Information obtainable on fluctuations in dust concentrations over short time periods	Cannot in general be used for total dust and those that can may not give readings which correlate with results obtained by gravimetric sampling
No weighing necessary, results can be obtained by non-technical personnel	Generally expensive (£3000–£6000 per instrument)

calibrate the instrument by comparison with the average dust concentration recorded on the instrument output (Simslin II monitor).

The second monitor uses a pulsed LED as the light source. Continuous readings are displayed in $mg\,m^{-3}$ units at an update rate of $3\,s^{-1}$. Time constants of 0·5, 2·8 and 32 s can be used to provide varying degrees of smoothing for the output. Unlike the Simslin, no memory facility is available on the monitor although an output for a strip chart recorder is provided for situations where mains electricity is available. The dust is not collected on a filter suitable for weighing in this instrument so that calibration by this means is not possible. A reference scatterer can however be inserted into the light path to check on the existing calibration. Forward scattering is intrinsically less affected by changes in particle size, composition, etc., than other forms of scattering, but if this instrument is to be used for different dusts it would be best to carry out a calibration check for each dust using a standard dust measuring technique. This is called the GCA–RAM–1 monitor.

These automatic samplers are certainly an advance over traditional pump-and-filter methods, but they do not at present provide a total answer to the hygienist's dust monitoring problems (see Table 16.1). They are therefore likely to be used in conjunction with, rather than in place of, the traditional methods.

FUME MEASUREMENT

'Fume' is a somewhat loose term but is usually used in the rubber industry to describe the blue smoke or haze generated by hot rubber compounds.

This fume is very complex. It contains small liquid droplets (aerosols), which make it visible, and invisible vapours.

The exact composition of both these phases depends on the materials present in the rubber compound, but it is usual for the vapour phase to contain dozens, and the aerosol phase hundreds, of components at the sub-ppm level. Some of the common vapours generated by rubber compounds are shown in Table 16.2.[4,5]

These materials may be monitored using the techniques given in the next section. It has been argued, however, that many of the materials capable of producing long term damage to health may be found in the aerosol phase. Analysis of this phase is extremely complicated and even when a detailed analysis is carried out it may be impossible to interpret because of lack of information on the hazards of individual components, or more especially on mixtures of these components. The British Rubber Manufacturers' Association have therefore suggested that a sensible approach is to limit the total concentration of this fume, and have recommended a limit of 0.25 mg m^{-3} for the cyclohexane-soluble portion of the aerosol phase of hot rubber fume.

TABLE 16.2
VAPOURS GENERATED BY RUBBER COMPOUNDS

Component	Likely source
By-products	
Acetone	
tert.-Butanol	Various peroxide cures
Acetophenone	
α,α-Dimethylbenzyl alcohol	
Carbon disulphide	Sulphur cures, especially thiuram and dithiocarbamate
Cyclohexylamine	CBS cures
Dibutylamine	
Diethylamine	Thiuram and dithiocarbamate cures
Dimethylamine	
Hydrogen sulphide	Sulphur cures, especially at high temperature
Morpholine	MBS and dimorpholinyl disulphide cures
Polymerisation residues	
Acrylonitrile	NBR
Chloroprene	CR
Divinylbenzene	IIR
Ethylidene norbornene	EPDM
Vinyl chloride	PVC

It is possible to measure this fraction of the fume using either static or personal sampling but because of the very small quantities of material collected on personal samples, great care is needed in the technique used. For instance a 3-h sample at 3·0 litres min^{-1} in an atmosphere containing 0·25 mg m^{-3} cyclohexane-soluble aerosol will capture only 135 μg cyclohexane-soluble material. Static samples can be collected using mains-operated samplers at about 15–25 litres min^{-1}; personal samples require battery operated samplers, but if possible modern samplers capable of sampling at 3–4 litres min^{-1} should be used to maximise the sample size. In both techniques the sample is collected on a glass fibre filter: before sampling a batch of filters should be exhaustively extracted (3 days) with cyclohexane in a Soxhlet before drying and allowing to equilibrate with the laboratory air. All loose fibres should be removed from the filter surface with a soft brush, and it should then be weighed, using tweezers for handling.

Because of the small weight of material which will be collected, a microbalance such as described for dust sampling is absolutely necessary.

Sampling must be carried out for sufficient time to produce a measurable weight of aerosol—at least 3–4 h for personal samples. After sampling, the filter and holder should be returned to the laboratory and allowed to equilibrate for at least 24 h before reweighing.

After weighing (using tweezers again) it should be folded in half with the sample inside. A conventional paper filter is then folded round the glass fibre filter and held in place with a paper clip. This is necessary to prevent any dust particles present from being dislodged and removed from the filter during the extraction procedure. Several folded 2·5 cm glass fibre filters can be placed in successive folds of an 11 cm diameter paper filter, the ends of which are then folded over to form an envelope and held with paper clips. The envelope is then extracted in a Soxhlet overnight. After extraction it is removed from the Soxhlet and allowed to dry in a gentle air stream. The glass fibre filters are then removed and allowed to equilibrate with the balance room atmosphere for at least $\frac{1}{2}$ h. They are then reweighed. The cyclohexane-soluble concentration is then calculated from the difference between second and third weighings. A blank filter, put through the whole procedure, can be used to allow for slight system weight changes. With careful technique this method will produce results from a 5 h sample in the range 0·05–1·0 mg m^{-3} cyclohexane-soluble material with an accuracy better than 20%.

The automatic dust monitors described previously will usually give a reading in areas contaminated by fumes, and may therefore be useful in assessing such areas. Where these instruments are used for this purpose, however, it should be borne in mind that the device used to remove coarse particles from the airstream in these instruments may affect the result given by fumes, as also may the operating system of the instrument itself.

GASES AND VAPOURS

There is a bewildering variety of equipment for measuring gases and vapours. It is possible to choose equipment for spot tests, time-weighted average measurements or for continuous monitoring, and any of these methods may be designed to give results on the spot, or after laboratory analysis. One requirement for almost all this equipment, however, is that it is specific for the material in question—it responds to that material and no other.

Spot or Grab Samples
The simplest form of grab sampling is carried out with an evacuated rigid container. The container is taken to the point where sampling is to be carried out and the tap opened. After the container has filled with the site atmosphere, the tap is closed and the container transported back to the laboratory for analysis, usually by gas chromatography. The container should be made of inert material so that it does not interact with the gas or vapour being measured. The system is best suited to sampling permanent gases.

A variant of this system is to use a small (100 ml) evacuated container fitted with a critical orifice. The size of the orifice controls the rate at which the air sample is drawn into the container, and with a suitable orifice the sampling time may be up to 8 h, giving a time-weighted average measurement (vacuum operated samplers—see Appendix).

Instead of evacuated containers, atmosphere samples may be obtained at the sampling site by flushing the air from a rigid container using a pump. The container is then sealed for subsequent analysis. All the above systems have the disadvantages that the sample size is restricted to the container size, which may limit the sensitivity of the measurement, and that they are not suitable for unstable or reactive materials.

Detector Tubes
These operate on the 'breathalyser' principle. They consist of glass tubes filled with a reactive chemical packing. Air is drawn through the tube,

usually by means of a hand pump, and in the presence of the specific contaminant, the packing gradually changes colour down the length of the tube. The colour boundary moves down the tube as the air sample is taken, and when a given volume of air has passed (usually 100–1000 ml), the boundary is read off against a scale printed on the tube. Detector tubes for more than 100 different materials are available from the major suppliers (see Appendix).

Detector tubes have normally been used to obtain measurements at a single place and time (spot or 'grab' sampling) but recently a limited number of tubes have been introduced for time-weighted average measurements. These use a low air flow rate provided by a battery operated pump over several hours to give the average concentration over this period. They can be used for personal sampling.

TWA Measurements—Pump and Trap Systems

Air may be pumped into an inflatable sampling bag made of inert material, which is then sealed for subsequent analysis. By using a low flow rate pump, a sample can be obtained over several hours to give a time-weighted average measurement. The system does however suffer from the disadvantages outlined for the equivalent grab sampling systems. It is now more usual to use a sampling pump to draw air samples through some form of trap, which collects the material in question and retains it for subsequent analysis. The following types are in common use:

Bubblers

These contain a small quantity of liquid through which the airstream is drawn. A sintered glass frit is often used to break up the airstream and ensure good contact with the absorbing liquid, which may be either a solvent (such as water or an organic solvent chosen to dissolve the airborne contaminant) or a reactive liquid system which will fix the contaminant chemically in a form suitable for subsequent analysis. Analysis of the trapped vapours may be made by physical methods (gas chromatography, high performance liquid chromatography, IR or UV spectroscopy, etc.) or by chemical methods. In the latter it is common to use a colorimetric method in which the contaminant is reacted with suitable reagents to give a coloured solution, the intensity of which is proportional to the original concentration of the contaminant. The intensity of the colour can be measured on site with a (Minispec) portable spectrophotometer or with rather less accuracy can be compared with coloured glass or liquid standards.

Many of these methods are detailed in the Health and Safety Executive's series of booklets '*Methods for the Detection of Toxic Substances in Air*', obtainable from HMSO,[6] and in the newer series *Methods for the Determination of Hazardous Substances.*[7]

Adsorber Tubes
These use a tube filled with adsorbent packing to trap the contaminant vapours. The most common adsorbent is activated charcoal, but other packings such as silica gel, Tenax and other chromatography packings are also used. The charcoal adsorber system, because of its wide applicability to most industrial solvents, monomers and many other volatile contaminants, has become one of the standard monitoring techniques. Having trapped the volatile contaminants on charcoal, the usual method of analysis is by gas chromatography. The contaminants are transferred from the charcoal to the chromatograph either by desorption into carbon disulphide (or other suitable solvent) and injection of the solution into the chromatograph or by flash heating of the charcoal and use of a gas sample loop to introduce the vapours produced to the chromatograph.

The charcoal trap/gas chromatograph system gives very good specificity and sensitivity, and can be used for the simultaneous measurement of several individual contaminants in, for example, mixed solvent systems. When using the method to give time-weighted average measurements on personal samples over periods greater than 1 h, there is a danger that the charcoal trap may be overloaded. This has led to the introduction of low flow rate personal sampling pumps designed to operate at air flows well below the normal range for conventional dust and other sampling. Flow rates in the range 0.5–$200\,\text{ml}\,\text{min}^{-1}$ are now standard.

Passive Samplers
Even modern sampling pumps require careful attention to flow rates, battery condition, etc., and are, however slightly, an encumbrance to the wearer. Passive samplers have been developed to obviate the necessity for the sampling pump. There are two main types:

Diffusion Based Samplers (Fig. 16.1)
These samplers contain an adsorptive layer at the base of the sampler and operate by diffusion of vapour molecules from the aperture of the sampler to this adsorptive layer. The aperture to the sampler is sometimes covered by a membrane which acts as a draft shield. This membrane offers no resistance to passage of vapour molecules.

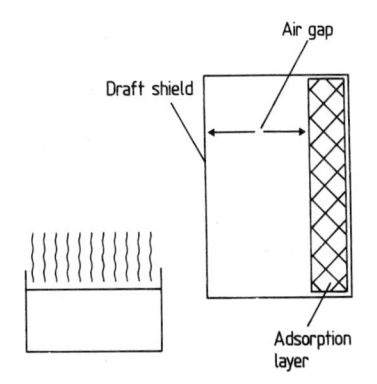

Fig. 16.1. Diffusion passive sampler.

The effective steady state rate of sampling of this type of passive sampler is given by Fick's first law:

$$N = \frac{DAC_\alpha}{\lambda}$$

where N = diffusive transport rate in mol s^{-1}, D = contaminant diffusivity in air in cm^2 s^{-1}, A = diffusion path cross-sectional area in cm^2, C_α = concentration of contaminant at the sampler aperture in mol cm^{-3} and λ = distance between aperture and adsorptive surface in cm. The equation assumes steady state sampling conditions and zero concentration of contaminant at the adsorptive surface.

It can be seen that for fixed values of D, A and λ, the rate of pollutant collection is directly proportional to the ambient concentration. The sampling rate for different vapours is usually supplied by the manufacturer of the passive sampler, or can be determined experimentally.

Permeation Based Samplers (Fig. 16.2)
These samplers again use an adsorptive layer of material, but this is now covered by a permeable membrane such as silicone rubber.

The permeable membrane controls the rate of absorption of vapour molecules by the adsorptive layer, the effective sampling rate being given by

$$F = \frac{PC_\alpha A}{d}$$

where F = permeation transport rate in cm^3 s^{-1}, P = permeability coefficient, C_α = concentration of contaminant at aperture of sampler,

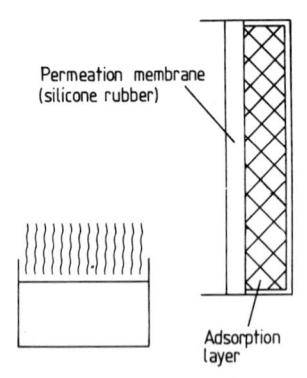

Fig. 16.2. Permeation passive sampler.

A = cross-sectional area of aperture, cm^2, d = thickness of permeable membrane, cm. Thus the effective rate of contaminant collection is proportional to the ambient concentration.

Most commercial samplers use either diffusion or permeation principles. The samplers take the form of badges, rather like the familiar radiation badge, which pin to the operator's lapel. After the sampling period, the amount of collected material is determined by soaking the adsorptive layer in a solvent, often carbon disulphide, and analysing the resulting solution by gas chromatography.

One recent passive sampler (Perkin Elmer model) uses a combination of permeation and diffusion principles, with a silicone rubber layer at the entrance to the sampler, and a diffusion gap between this and the adsorption layer. The sampler is in the form of a small tube with a metal body, and is designed to use thermal desorption to remove the collected vapours. After sampling the tube is fitted into an electrically heated desorber where the collected material is rapidly vaporised and flushed into the gas chromatograph. An automated system for sequential loading and analysis of 50 sample tubes is available.

Although passive samplers do simplify the sampling procedures, it is necessary to take some care that they are suitable for the problem in hand. The early samplers suffered from variable sampling rates with different air velocities and directions across the sampler face. Changes in humidity and temperature have also been found to cause changes in sampling rate for some materials. The samplers also work best in relatively steady concentrations of contaminant, and the results obtained for the average concentration may be in error in situations where rapid fluctuations in concentration are present. It is therefore best to carry out a check on the accuracy of these samplers by comparison with a standard sampling method, usually an adsorbent tube and pump system, before they are used in routine sampling work. In spite of these drawbacks, passive samplers can be extremely useful, particularly in situations where routine samples are required at locations when no trained hygienist is available to carry out more conventional sampling.

Continuous Gas and Vapour Monitors
Sensitised Paper Tapes
In the simplest form of tape sampling, air is drawn through a chemically treated test paper to produce a stain in the presence of a specific atmospheric contaminant. The stain can then be compared with standards or measured instrumentally to give an assessment of the concentration of the contaminant. The technique has, however, been considerably developed in recent years, and now represents one of the most advanced forms of monitoring.

The first development was to use a continuously moving strip of tape (continuous tape monitor) operating in the manner of a tape recorder. As the tape passes through the sampling head, air is drawn through it to give a stain in the form of a continuous streak of colour of varying intensity. This stain then passes through an optical reader positioned just after the sampling point where an electrical signal proportional to the stain intensity is produced. The signal is then used to give a meter reading calibrated in terms of concentration of the particular contaminant or can be fed to a chart recorder. The monitor thus provides a continuous reading of the changing atmospheric concentration of the contaminant which can be presented as a concentration versus time graph.

The next development was to miniaturise this system (MCM tape monitor) so that it could be used as a personal sampler. The main difficulty here was to make the optical system small enough to fit into a personal sampler, and in the present form of the miniaturised continuous monitor, this problem is sidestepped by leaving the optics out of the sampler. The monitoring head thus contains the tape, tape drive and air pump system and produces an 8-h tape. At the end of the shift the tape is removed from the monitor and fed through a separate reader/recorder containing the optical measuring system.

The miniaturised continuous monitor (MCM) is the first system by which continuous readings of the peaks and troughs of an individual's exposure to a contaminant may be obtained, as opposed to a simple time-weighted average. As such it produces the most detailed information of exposure which is currently available, and points the way to future monitoring systems. The present system still has some drawbacks, the main one being that it can only be used with tapes giving a stable stain.

The advent of the microprocessor has opened the possibility of producing a miniaturised sampling head including an optical measuring system, and developments are under way to produce a personal tape monitor of this kind, which will also incorporate signal storage facilities. This system will enable a wider range of tapes to be used. In this sampler it is likely that the tape will be driven in discrete steps through the sampling head, this method giving advantages in terms of resolution.

Meanwhile, the tape system has also been considerably developed for use in spot sampling—that is, the production of a single reading over a short measuring period. The latest instrument (Autospot 3060 monitor) operates using cards containing tape cassettes and is controlled by a microprocessor to give the correct sampling time, flow rate and optical measurement system for the tape in use. The reading for concentration of the contaminant during the test is displayed on an electronic bar display. Tapes for some 12 different atmospheric contaminants are available for this instrument.

Optical Monitors

Many of the functional groups present in organic compounds absorb infrared radiation strongly at specific wavelengths. Miran portable infrared spectrometers utilising this principle have been introduced and are capable of giving instantaneous continuous readings for organic vapours of this kind. In the normal version of this instrument, air is drawn through the instrument cell by means of a small pump. To obtain the required sensitivity at the low concentrations found for many industrial contaminants, a long path length is necessary. This is obtained by using multiple reflections of the infrared beam in the 0·75 m long gas cell to give a final optical path length of up to 20 m. For a specific contaminant, the instrument is set to provide an infrared beam at the wavelength at which this contaminant gives its maximum absorption. The spectrometer then measures the absorption of energy of occurring after passage of the beam through the gas cell, and this measurement is used to obtain a reading for the concentration of this contaminant in the atmosphere.

The instrument gives a rapid on-the-spot measurement for many organic vapours, and can be used to plot out the contours of concentration of materials in the vicinity of industrial processes. It is essentially an 'area' method, however, and is too large for use as a personal sampler. Also, although the instrument possesses a reasonable degree of specificity, it will respond to any compound absorbing at the set wavelength, and may therefore be subject to interference under some circumstances.

Electrochemical Monitors and Dosimeters

These monitors use an electrochemical cell system designed to produce an electric current by the oxidation or reduction of a specific atmospheric contaminant in the cell. A diffusion membrane is often used to control the transfer of the contaminant from atmosphere to the cell. These monitors may use the current generated by the cell to operate a directly calibrated meter, or they may incorporate a battery operated alarm system triggered at a pre-set level by the cell output. Personal monitors using this system have been developed and in some cases these measure the total exposure to the contaminant by recording the total cell current output (dosimeters). Monitors for oxygen, carbon monoxide, hydrogen sulphide, oxides of nitrogen, ammonia and phosgene are available.

It should be noted that the working lifetimes of these cells are finite, and that the cell itself will therefore require replacement at intervals. Calibration with known gas concentrations is also necessary at regular intervals.

Solid State Monitors

It has been found that the presence of many atmospheric contaminants will change the electrical conductance of semi-conductor materials, and this phenomenon has been utilised to produce the 'solid state monitors'. A voltage is applied to the semi-conductor, and the steady current produced in the absence of atmospheric contaminants is used to set the instrument zero. If atmospheric contaminants are now allowed to diffuse through to the semi-conductor, a change in conductance is produced which is proportional to the concentration of the contaminant. The signal current through the semi-conductor is used to give a direct reading of the concentration of atmospheric contaminant.

These monitors are essentially non-specific, and will give readings in the presence of many atmospheric contaminants. To some extent they can be made more sensitive to particular contaminants by choice of semi-conductor material or operating voltage, but interference by other contaminants is impossible to eliminate completely. They have been most used in hydrogen sulphide monitors.

Portable Gas Chromatographs
The gas chromatograph is an extremely useful analytical tool for measurement of volatile materials, since when used with the now usual flame ionisation detector it is a sensitive and reasonably specific instrument. The standard laboratory instrument, with its associated gas cylinders and electrical suppliers is much too large for use as a portable factory floor monitor. Much ingenuity has however been used to develop battery operated portable gas chromatographic equipment, and models are now available for use in this fashion.

The key part of these monitors is the flame ionisation (FI) detector. When organic molecules enter a hydrogen flame they are ionised and cause the flame to be electrically conducting. By measuring this conductivity the concentration of the organic material can be estimated.

In these portable chromatographs, the FI detector can be used on its own, in which case a reading is obtained for the total quantity of organic material in the atmosphere, or the air sample can be switched through a chromatograph column so that the individual organic components are measured in sequence. As with the infrared spectrophotometer the instruments can be used to plot the concentrations of contaminant(s) round the workplace.

Several materials can be measured simultaneously when using this type of instrument in its gas chromatography mode.

These instruments are particularly suited for monitoring materials such as mixed solvent vapours. It may be noted, however, that although the instruments are truly portable, they are still too large to use as personal monitors. Compared with laboratory-based gas chromatographs, they are also usually limited by the need to operate the chromatography column at room temperature or at fixed temperatures close to this.

CONCLUSIONS

The monitoring equipment field is developing rapidly, and many new instruments or modifications to existing systems will make their appearance in the next few years. It seems likely that many of these developments will concentrate on equipment for personal sampling since experience has shown that results obtained in this manner are very often considerably different from the results obtained by static or area monitors. It is, after all, the person's real exposure under practical working conditions that needs to be measured.

Also, it is probable that continuous monitors will become more important than grab or spot samplers, and that some of these continuous monitors will produce information on fluctuations in concentrations with time, as well as time-weighted average measurements.

We have already seen the trend towards instruments which produce on-the-spot readings, and this is likely to grow. Paradoxically, however, the collection of samples by simple methods for subsequent laboratory analysis is also likely to grow as ever more sophisticated analytical methods are required to measure the tiny traces of contaminants now being considered as significant.

REFERENCES

1. *Threshold Limit Values for 1980*, Guidance Note EH15/80, Health and Safety Executive, London, 1980.
2. *Documentation to the Threshold Limit Values*, American Conference of Governmental Industrial Hygienists, Cincinnati, 1982.
3. *Toxicity and Safe Handling of Rubber Chemicals*, Code of Practice, British Rubber Manufacturers' Association, Birmingham, 1978.
4. Ashness, K., Lawson, G. and Willoughby, B. G., *Plastics and Rubber Institute Rubber Conference*, Harrogate, 1981.
5. Willoughby, B. G., *Plastics and Rubber Institute General Rubber Goods Technology Conference*, Sutton Coldfield, 1982.
6. Health and Safety Executive, *Methods for the Detection of Toxic Substances in Air*, 26 booklets available from HMSO, London.
7. *Methods for the Detection of Hazardous Substances*, Guidance Notes MDHS 1–30, Health and Safety Executive, Occupational Medicine and Hygiene Laboratories, London.

Appendix

SOURCES OF EQUIPMENT MENTIONED IN CHAPTER 16

Item	Company	Address
Autospot 3060 monitor	MDA Scientific Inc.	Park Ridge, Illinois, USA
	MDA Scientific (UK) Ltd	Wimborne, Dorset, UK
Continuous tape monitors	MDA Scientific Inc.	Park Ridge, Illinois, USA
	MDA Scientific (UK) Ltd	Wimborne, Dorset, UK
Detector tubes		
Bendix-Gastec tube	National Environmental Instruments Inc.	Warwick, Rhode Island, USA
Draeger tube	Drägerwerk	Lübeck, West Germany
	National Draeger Inc.	Pittsburgh, Pennsylvania, USA
	Draeger Ltd	Chesham, Buckinghamshire, UK
Kitegawa tube	Komyo Chemical Co.	Tokyo, Japan
MSA tube	Mine Safety Appliances Co.	Pittsburgh, Pennsylvania, USA
Siebetec tube	Siebe Gorman Ltd	Cwmbran, Gwent, UK
GCA-RAM-1 monitor	GCA Environmental Instruments	Bedford, Massachussetts, USA
	Analysis Automation Ltd	Oxford, UK
GCA-RDM automatic dust monitors	GCA Technology Division	Bedford, Massachusetts, USA
	Analysis Automation Ltd	Oxford, UK
MCM tape monitor	MDA Scientific Inc.	Park Ridge, Illinois, USA
	MDA Scientific (UK) Ltd	Wimborne, Dorset, UK
Minispec portable spectrophotometer	Bausch and Lomb	Rochester, New York, USA
	Gallenkamp Ltd	Loughborough, Leicestershire, UK
Miran infrared portable spectrophotometer	Wilks Scientific Corp.	South Norwalk, Connecticut, USA
	Foxboro Analytical Ltd	Milton Keynes, Buckinghamshire, UK
Modified UKAEA holder	C. F. Casella & Co. Ltd	London, UK

APPENDIX—contd.

Item	Company	Address
Passive samplers		
Abcor Gasbadge	Abcor Development Corp.	Wilmington, Massachussetts, USA
3M's organic vapour monitor	3M Occupational Health and Safety Products	Bracknell, Buckinghamshire, UK
Orsa 5 samplers	Drägerwerk	Lübeck, West Germany
Perkin-Elmer model ATD50	Perkin-Elmer Corp.	Norwalk, Connecticut, USA
	Perkin-Elmer Ltd	Beaconsfield, Buckinghamshire, UK
Protec organic vapour monitoring badges	E. I. DuPont de Nemours & Co. Inc.	Wilmington, Delaware, UK
	Shawcity Ltd	Farringdon, Oxon, UK
Portable gas chromatographs		
Century organic vapour analyser	Century Systems Corp.	Arkansas City, Kansas, USA
	Foxboro Analytical Ltd	Milton Keynes, Buckinghamshire, UK
Unico PGC portable gas chromatograph	National Environmental Instruments Inc.	Avondale, Pennsylvania, USA
Model 511 portable gas chromatograph	Analytical Instrument Development Inc.	Avondale, Pennsylvania, USA
	Vinten Instruments	Weybridge, Surrey, UK
Simslin II monitor	Rotheroe & Mitchell	Ruislip, Middlesex, UK
TSI-3500 automatic dust monitor	Thermosystems Inc.	St. Paul, Minnesota, USA
	Bristol Industrial and Research Assoc. Ltd	Portishead, Bristol, UK
Vacuum operated samplers	C. F. Casella & Co. Ltd	London, UK

Bibliography

CARCINOGENICITY, MUTAGENICITY, TERATOGENICITY

Clayson, D. B., *Chemical Carcinogenesis*, Churchill, London, 1962.
De Serres, F. J. and Ashby, J., Evaluation of short-term tests for carcinogens, *Progress in Mutation Research*, Vol. 1, Elsevier/North-Holland, Amsterdam, 1981.
Fishbein, L., *Potential Industrial Carcinogens and Mutagens*, Elsevier, Amsterdam, 1979.
GMWU/City University, *Cancer and Work: Making Sense of Workers Experience*, London, 1982.
Hueper, W. C. and Conway, W. D., *Chemical Carcinogenesis and Cancers*, Charles C. Thomas, Springfield, Illinois, 1964.
Kimmel, C. A. and Buelke-Sam, J., *Development Toxicology*, Raven Press, New York, 1981.
Sax, N. I., *Cancer-causing Chemicals*, Van Nostrand Reinhold, New York, 1981.
Scott, D., Bridges, B. A. and Sobels, F. H., *Progress in Genetic Toxicology*, Elsevier/North-Holland, Amsterdam, 1977.
Searle, C. E., *Chemical Carcinogens*, ACS Monograph 173, American Chemical Society, Washington DC, 1976.
Shepard, T. H., *Catalogue of Teratogenic Agents*, 3rd Edn, Johns Hopkins University Press, Baltimore, 1980.
Wilson, J. G. and Fraser, F. C., *Handbook of Teratology*, Vols 1 and 2, Plenum Press, New York, 1977 and 1979.

GENERAL AND INDUSTRIAL TOXICOLOGY

Aldridge, W. N., *Mechanisms of Toxicity*, Macmillan, London, 1968.
Anderson, K. and Scott, R., *Fundamentals of Industrial Toxicology*, Ann Arbor Science, Ann Arbor, Michigan, 1981.

Bretherick, L., *Handbook of Reactive Chemicals Hazards*, 2nd Edn, Butterworths, London, 1979.
Cassarett, L. J. and Doull, J., *Toxicology: The Basic Science of Poisons*, 2nd Edn, Macmillan, London, 1980.
Daugaard, J., *Symptoms and Signs in Occupational Disease—A Practical Guide*, Munksgaard, Copenhagen, 1978.
Gosselin, R. E., Hodge, H. C., Smith, R. P. and Gleason, M. N., *Clinical Toxicology of Commercial Products*, 4th Edn, Williams & Wilkins, Baltimore, 1976.
Hunter, D., *Diseases of Occupation*, 6th Edn, Hodder & Stoughton, London, 1978.
Loomis, T. A., *Essentials of Toxicology*, 3rd Edn, Lea & Febiger, Philadelphia, 1978.
Patty's Industrial Hygiene and Toxicology, 3rd Revised Edn, Wiley-Interscience, New York, 1982.
Plunkett, E. R., *Handbook of Industrial Toxicology*, Heyden, New York, 1976.
Sax, N. I., *Dangerous Properties of Industrial Materials*, 5th Edn, Van Nostrand, New York, 1979.

LIMIT VALUES

American Conference of Governmental Industrial Hygienists, *Documentation of the Threshold Limit Values*, Cincinnati, Ohio, 1982.
American Conference of Governmental Industrial Hygienists, *Threshold Limit Values for Chemical Substances and Physical Agents in the Work Environment, with Intended Changes for 1983/84*, Cincinnati, Ohio, 1982.
Deutsche Forschungsgemeinschaft, *Maximalearbeitsplatzkonzentrationen (MAK) List*, Bonn, 1982.
Health and Safety Executive, *Threshold Limit Values for 1980*, HSE Guidance Note EH15/80, HMSO, London, 1980.

PHYSIOLOGICAL EFFECTS

Cronin, E., *Contact Dermatitis*, Churchill Livingstone, London, 1980.
Fregbert, S., *Occupational Contact Dermatitis*, 2nd Edn, Year Book Medical Publishers Inc, Chicago, 1981.

Grant, W. M., *Toxicology of the Eye*, Charles C. Thomas, Springfield, Illinois, 1974.
Lee, D. H. K., Falk, W. L., Murphy, S. D. and Geiger, S. R., *Handbook of Physiology, Section 9: Reactions to Environmental Agents*, American Physiological Society, New York, 1977.
Spencer, P. S. and Schaumberg, H. H., *Experimental and Clinical Neurotoxicology*, Williams & Wilkins, Baltimore, 1980.
Weiss, B. and Latis, V. G., *Behavioural Toxicology*, Plenum Press, New York, 1975.
Williams, R. T., *Detoxification Mechanisms*, 2nd Edn, Chapman and Hall, London, 1959.

RUBBER INDUSTRY

British Rubber Manufacturers' Association, *Toxicity and Safe Handling of Rubber Chemicals: Code of Practice*, Birmingham, 1978.
Holmberg, B. and Sjöström, B., *A Toxicological Survey of Chemicals used in the Swedish Rubber Industry*, Unit of Occupational Toxicology, National Board of Occupational Health and Safety, Stockholm, 1977.
International Agency for Research on Cancer, *Monograph on the Evaluation of the Carcinogenic Risk of Chemicals to Humans—Vol. 28: The Rubber Industry*, Lyon, France, 1982.
Malten, K. E. and Zeilhuis, R. L., *Industrial Toxicology in the Production and Processing of Plastics*, Elsevier, New York, 1964.
Olsson, S., *Proposals for the Elimination of Chemical Hazards in the Rubber Industry*, Unit of Occupational Toxicology, National Board of Occupational Health and Safety, Stockholm, 1977.
Rubber and Plastics Research Association, *Clearing the Air—A Guide to Controlling Dust & Fume Hazards in the Rubber Industry*, Shrewsbury, UK, 1982.

TOXICITY OF SPECIFIC GROUPS OF CHEMICALS

Browning, E., *Toxicity and Metabolism of Industrial Solvents*, Elsevier, Amsterdam, 1965.
Jones, P. W. and Leber, P. (Eds), *Polynuclear Aromatic Hydrocarbons*, Ann Arbor Science, Ann Arbor, Michigan, 1979.

Medical Research Council, *The Carcinogenic Action of Mineral Oils: A Chemical and Biological Study*, MRC SRS 306, HMSO, London, 1968.

National Research Council/National Academy of Sciences, *Particulate Polycyclic Organic Matter*, Washington DC, 1972.

Scott, T. S., *Carcinogenic and Chronic Toxicity Hazards of Aromatic Amines*, Elsevier, Amsterdam, 1962.

Von Oettingen, W. F., *The Halogenated Hydrocarbons of Industrial and Toxicological Importance*, Elsevier, Amsterdam, 1964.

TOXICOLOGICAL TESTING

European Economic Community, *Directive 79/831/EEC, Annex V*, 18 September 1979.

Her Majesty's Government, *Notification of New Substances Regulations 1982, Approved Code of Practice—Methods for the Determination of Toxicity*, HMSO, London, 1982.

World Health Organisation, *Principles and Methods for Evaluating the Toxicity of Chemicals, Part I: Environmental Health Criteria*, Geneva, 1978.

Chemical Trade Names Index

A1, 88
Agerite
  Resin, 31
  White, 33
Altax, 84
Altofane A, 101
Ancazate EPH, 76
Anchor
  ADPA, 99
  CBS, 81
  DNPD, 107
  DPG, 79
  DPPD, 106
  HDPA, 108
  IPPD, 104
  MBT, 83
  MBTS, 84
  PBN, 102
  SPH, 109
  TMQ, 100
  TMTD, 90
  TMTM, 89
  ZDBC, 75
  ZDEC, 75
  ZDMC, 74
  ZMBT, 84
Antioxidant
  2246, 111
  AH, 31
  AP, 31
  DNP, 33
Antioxygene RA, 31
Aranox, 106
Azobul, 115

Bantex, 84
Beutene, 72
BLE 25, 99
Butazate, 75

Captax, 83
Cellosolve, 125
  acetate, 125
Celogen
  AZ, 115
  BSH, 116
  OT, 117
Cereclor, 69
Cumate, 77
Curetard A, 96

Delac
  MOR, 82
  NS, 81
  5, 81
Diak
  No. 1, 66
  No. 3, 66
Dibenzo GMF, 66
Dicup
  40C, 64
  KE, 64
  R, 64
  T, 64
Dipentax, 116
DOTG, 79
DPG, 79

Ekagom
  CBC, 81
  D, 79
  4R, 75
Ethasan, 75
Ethazate, 75
Ethyl
  Cadmate, 78
  Tuex, 91
  Zimate, 75

Flectol
  flakes, 100
  pastilles, 100
Flexone
  3C, 104
  18F, 104
  4L, 103
  7L, 105
  8L, 104

Genitron
  AC, 115
  AZDN, 117
  BSH, 116
  OB, 117

HDPA, 108
Heptene base, 72
Hexa, 73
HMT, 73

J-2-F, 106

Lorvinox
  44S36, 113
  44S36P, 113
  ACP, 100
Luperco
  540C, 64
  540KE, 64
  500R, 64
  500T, 64
Luperox, 65

Methasan, 74
Methazate, 74
Methyl
  Cellosolve, 125
  acetate, 125
  Oxitol, 125
  acetate, 125
Monex, 89
Monothiurad, 89
Montaclere, 109

Naugard
  BHT, 110
  NBC, 77
  Q, 100
  SP, 109
  431, 111
  445, 109
Naugawhite, 111
NOBS special, 82
Nonox S, 6, 7, 13, 15, 25, 26, 29–31
  composition, 29–30
  β-naphthylamine content, 30
  β-naphthylamine in atmosphere, 30–1

Octamine, 108
Oxaf, 84
Oxitol, 125
  acetate, 125

Perkadox
  BC-90, 64
  BC-40B, 64
  BC-40K, 64
  14-40K, 65
  SB, 64
  14-90, 65
Permalux, 80
Permanax
  B (pastilles), 99
  BL, 99
  BLN, 99
  BLW, 99
  DPPD, 106

CHEMICAL TRADE NAMES INDEX

Permanax—*contd.*
  EXP, 100
  HD, 108
  IPPD, 104
  OD, 108
  6PPD, 105
  SP(L), 109
  TQ, 100
  WSL, 110
  WSO, 113
  WSP, 112
Polylite, 108
Porofor
  ADC/R, 115
  BSH, 116
  DWO, 116

Retarder
  ESEN, 98
  J, 45, 96
  PD, 98
  TSA, 97
Rhenocure CA, 88
Rhenogran DOTG, 79
Robac
  CuDD, 77
  DETU
    flake, 87
    PM, 87
    90, 87
  MX-1, 84
  NiBUD, 77
  PPD, 78
  PTD, 92
  TBTU, 92
  TET, 11
  Thiuram, P25, 93
  TMS, 89
  TMT, 90
  Tribtu, 88
  ZBED, 76
  ZBUD, 75
  ZDC, 75
  ZIX, 93
  ZMD, 74
  ZPD, 76
  22, 85

Santocure, 81
  MOR, 82
  NS, 81
Santoflex
  AW, 101
  B, 27, 34
  BX, 34
  IP, 104
  13, 105
  77, 103
Santogard PVI, 97
Santowhite
  Crystals, 113
  Powder, 112
SBP
  1, 121
  2, 121
  3, 121
Sulfasan R, 63

Thiofide MBTS, 84
Thiotax, 83
Thiurad, 90
Thiuram M, 90
Triganox
  B, 64
  C, 65
  C50D, 65
Trimene base, 73

Unicel ND, 116
UOP
  26, 104
  288, 103
  588, 105
  688, 105
  788, 103
  88, 104
Ureka base, 85

Vanguard N, 77
Vazo, 64, 117
Vocol, 94
Vulcacel BN94, 116

Vulcafor
  CBS, 81
  DCBS, 82
  DOTG, 79
  DPG, 79
  MBS, 82
  MBT, 83
  MBTS, 84
  TMTD, 90
  TMTM, 89
  ZDBC, 75
  ZDEC, 75
  ZDMC, 74
  ZMBT, 84
Vulcatard
  A, 45, 96
  SA, 97
Vulcastad EFA, 73
Vulkacit
  BZ/C, 82
  1000/C, 80
  CRV, 93
  CZ, 81
  D, 79
  DM, 84
  DOTG/C, 79
  H30, 73
  L, 74

Vulkacit—*contd.*
  LDA, 75
  Merkapto, 83
  MOZ, 82
  NN/C, 85
  NZ, 81
  P, 78
    extra N, 76
  Thiuram, 90
  Thiuram MS, 89
  ZMBT, 84
  ZP, 76
Vulkalent A, 45, 96
Vulkanox
  BKF, 111
  HS, 100
  KB, 110
  PAN, 101
  PBN, 102
  3100, 107
  4020, 105
  4030, 103

White spirit, 121
Wingstay
  S, 109
  100, 107

Chemical Names Index

Acetone, 27, 123, 164
Acetone/4-aminodiphenyl condensation product, 34
Acetone/diphenylamine condensation product, 34, 99
Acetophenone, 64, 164
Acrylonitrile, 53, 54, 164
Adipate esters, 71
Aldol α-naphthylamine, 31
Aluminium hydroxide, 59
Aluminium silicate, 59
4-Aminodiphenyl, 27, 34, 99
Amyl acetate, 124
Aniline, 32
Anthanthrene, 36, 38, 43
Aromatic oils, 35–7, 54
Asbestos, 58, 59
Azobis-isobutyronitrile, 117
Azodicarbonamide, 115

Barium sulphate, 59
Benz[a]anthracene, 36, 43
Benz[a]fluorene, 36, 43
Benzene, 18, 120, 135
Benzenesulphonyl hydrazide, 116
Benzidene, 24, 25, 34
Benzo[g,h,i]fluoranthene, 38, 43
Benzo[g,h,i]perylene, 36, 38, 43
Benzo[a]pyrene, 35–44, 69
Benzo[e]pyrene, 35, 36, 38, 43
Butadiene, 54
tert. Butanol, 164
n-Butyl acetate, 124
sec.-Butyl acetate, 124
tert.-Butyl acetate, 124
N-tert.-Butyl-2-benzothiazyl sulphenamide, 80–1
4,4′-Butylidene-bis-(6-tert.-butyl-m-cresol), 112
tert. Butylperoxybenzoate, 65
bis (tert.-Butylperoxyisopropyl) benzene, 65
Butyraldehyde/aniline condensation products, 72

Cadmium diethyldithiocarbamate, 78
Calcium carbonate, 59
Calcium silicate, 59
Carbazoles, 44
Carbon black, 10, 37–44, 142
Carbon disulphide, 135, 164
Carbon monoxide, 58, 122
Carbon tetrachloride, 54, 122, 135
China clay, 59
Chlorinated petroleum waxes, 69
Chloroprene, 53–4, 164
Chlorosulphonated polyethylene, 54
Chrysene, 36, 43
Clays, 59, 60
Copper dimethyldithiocarbamate, 77
Coronene, 38
Coumarone–indene resins, 69
Cyclohexylamine, 164
N-Cyclohexyl-2-benzothiazyl sulphenamide, 81
Cyclohexylthiophthalimide, 97
Cyclopenta[c,d]pyrene, 38

CHEMICAL NAMES INDEX

Dialphanyl phthalate, 70
Dianisidine, 34
Diaryl-*p*-phenylenediamines, mixed, 107
Diatomaceous earth, 60
Dibenz[*a,h*]anthracene, 35, 36, 43
Dibenz[*a,j*]anthracene, 36, 43
2,2′-Dibenzothiazyl disulphide, 47, 84
1,4-Dibenzoyl-*p*-benzoquinone dioxime, 66
1,2-Dibromo-3-chloropropane, 139
Dibutylamine, 164
2,6-Ditert.-butyl-*p*-cresol, 110
Di(5-tert.-butyl-4-hydroxy-2-methylphenyl) sulphide, 113
Ditert.-butyl peroxide, 64
Dibutyl phthalate, 70
Diisobutyl phthalate, 70
Dibutyl tin dilaurate, 55
Dichlorobenzidiene, 34
Dichloromethane, 122
N,N′-Dicinnamylidene-1,6-hexanediamine, 66
Dicumylperoxide, 64
N,N′-Dicyclohexyl-2-benzothiazyl sulphenamide, 82
N,N′-Dicyclohexyl-*p*-phenylenediamine, 104
Diethylamine, 45, 164
Di-2-ethylhexyl adipate, 71
Di-2-ethylhexyl phthalate, 70
Di-2-ethylhexyl sebacate, 71
1,3-Diethylthiourea, 87
4,4′-Diheptyldiphenylamine, 108
Dimethylamine, 45, 46, 164
α,α-Dimethylbenzyl alcohol, 64, 164
N-1,3-Dimethylbutyl-*N*′-phenyl-*p*-phenylenediamine, 105
2,4-Dimethyl-6-(1-methylcyclohexyl)-phenol, 110
2,6-Dimethylnitrosopiperidine, 44
N,N′-bis(1,4-Dimethylpentyl)-*p*-phenylenediamine, 103
N,N′-Di-β-naphthyl-*p*-phenylenediamine, 33–4, 107
2-(2′,4′-Dinitrophenylthio) benzothiazole, 85
Dinitrosopentamethylenetetramine, 116

Dioctyl adipate, 71
Dioctyl phthalate, 70
Dipentamethylenethiuram disulphide, 92
Dipentamethylenethiuram tetrasulphide/hexasulphide, 92–3
Diphenylamine/α-methylstyrene reaction product, 109
N,N′-Diphenyl guanidine, 79
Diphenylnitrosamine, 45
N,N′-Diphenyl-*p*-phenylenediamine, 106
1,3-Diphenyl-2-thiourea, 88
Distyrenated *p*-cresol, 111
Distyrenated hydroxytoluene, 111
Dithiodimorpholine, 63
Di-*o*-tolyl guanidine, 79
Di-*o*-tolyl guanidine salt of dicatechol borate, 80
Divinylbenzene, 54, 164

Ebonite, 15
EPDM rubbers, 54
Ethanol, 123
2-Ethoxyethanol, 125
2-Ethoxyethylacetate, 125
6-Ethoxy-2,2,4-trimethyl-1,2-dihydroquinoline, 101
Ethyl acetate, 124
Ethyl alcohol, 123
Ethylene glycol
  ethers, 124–7
  monoethyl ether, 125
  monoethyl ether acetate, 125
  monomethyl ether, 125
  monomethyl ether acetate, 125
Ethylenethiourea, 85
2-Ethylhexyl-diphenyl-phosphate, 71
Ethylidene norbornene, 54, 164
N,N′-bis(1-Ethyl-3-methylpentyl)-*p*-phenylenediamine, 104

Fluoranthene, 36, 38, 43
Formaldehyde/ammonia/ethyl chloride condensation product, 73

Glycol ethers, 124–7, 135

CHEMICAL NAMES INDEX 189

Heptaldehyde/aniline condensation products, 72
n-Heptane, 121
Hexamethylenediamine carbamate, 66
Hexamethylene tetramine, 73
n-Hexane, 121
2,5-Hexanedione, 121, 123
Hexane, other isomers, 121
Hydrogen cyanide, 135
Hydrogen sulphide, 164

Indeno[1,2,3-c,d]pyrene, 36, 38, 43
Isoprene, 54
N-Isopropyl-N'-phenyl-p-phenylenediamine, 104

Kieselguhr, 60

Lead, 135, 139
Lithopone, 59

Magnesium oxide, 59
2-Mercaptobenzothiazole, 83
2-Mercaptoimidazoline, 85
Methanol, 124
2-Methoxyethanol, 125
2-Methoxyethylacetate, 125
Methyl acetate, 124
Methyl alcohol, 124
Methyl n-butyl ketone, 123
Methylene chloride, 122
2,2'-Methylene-bis (4-methyl-6-tert.-butylphenol), 111
2,2'-Methylene-bis (6-[1-methylcyclohexyl]-p-cresol, 112
2,2'-Methylene-bis (4-methyl-6-nonylphenol), 111
Methyl ethyl ketone, 123
N,N'-bis(1-Methylheptyl)-p-phenylenediamine, 103
Methyl isobutyl ketone, 123
2-Methylnitrosopiperidine, 44
3-Methyl-2-thione-thiazolidine, 93
Mica, 59

Mineral fibres, man-made, 158
Morpholine, 46, 164
2-Morpholinothiobenzothiazole, 82

β-Naphthol, 32, 33
α-Naphthylamine, 25, 29, 30, 31, 34
β-Naphthylamine, 24, 25, 29–33, 54, 101, 102
2-Naphthylhydroxylamine, 33
Nickel dibutyldithiocarbamate, 77
4-Nitrodiphenyl, 34
Nitrogen oxides, 46
N-Nitrosamines, 19, 44–8
N-Nitrosodiethylamine, 45, 46, 47
N-Nitrosodimethylamine, 45, 46, 47
N-Nitrosodiphenylamine, 45–8, 96
N-Nitrosomorpholine, 46, 47
N-Nitrosopiperidine, 44, 46
N-Nitrosopyrrolidine, 46
Nonane, 121
Nonylated diphenylamine, 108
2,2'-Nonylene-bis(4,6-dimethylphenol), 113

Octane, 121
Octylated diphenylamine, 108
Oils, mineral, 68
p,p'-Oxy-bis (benzenesulphonyl hydrazide), 117
Ozone, 135

n-Pentane, 121
Perylene, 43
Petroleum
 naphtha, 36
 resins, 69
 rubber solvent, 121
 waxes, 68
N-Phenyl-N'-1-methylheptyl-p-phenylenediamine, 105
Phenyl α-naphthylamine, 101
Phenyl β-naphthylamine, 32–3, 54, 102
N-Phenyl-N'-(p-toluenesulphonyl)-p-phenylenediamine, 106
Phosgene, 135

Phosphate esters, 70
Phthalate esters, 70
Phthalic anhydride, 98
Phthalimide, 44, 97
Piperidinium pentamethylene dithiocarbamate, 78
Polycyclic aromatic hydrocarbons, 19, 35–44, 57–8
iso-Propanol, 124
iso-Propyl acetate, 124
n-Propyl acetate, 124
Pyrene, 35, 36, 38, 43

Quartz, 59–60

Resorcinol, 60–1
Rubber solvent, 121

Salicylic acid, 97
SBP solvents, 121
Sebacate esters, 71
Secondary amines, 45
see also under nitrosamines
Silica
  amorphous, 60
  crystalline, 59–60, 135, 142
Silicon tetrachloride, 60
Silicone polymers, 54
Sodium silicate, 60
Styrenated phenols, 109
Sulphur, 63

Talc, 8, 9, 10, 19, 58–9, 142
Tetrabutylthiuram disulphide, 92
Tetraethylthiuram disulphide, 45, 91
Tetramethylsuccinonitrile, 117
Tetramethylthiuram disulphide, 45, 47, 90
Tetramethylthiuram monosulphide, 89

4,4'-Thio-bis(6-tert.-butyl-*m*-cresol), 113
Thiocarbanilide, 88
Titanium oxide, 59
o-Tolidine, 34
Toluene, 120
o-Tolyl biguanidide, 80
1,1,3-Tributylthiourea, 88
1,1,1,-Trichloroethane, 122
Trichloroethylene, 122
Tricresyl phosphate, 70
2,2,4-Trimethyl-1,2-dihydroquinoline, polymerised, 100
2,2,4-Trimethyl-6-phenyl-1,2-dihydroquinoline, 34
Triphenyl phosphate, 70
Tritolyl phosphate, 70
Trixylyl phosphate, 70

Vinyl chloride, 54, 164

White spirit, 121

Xylene, 120
Xylenol–aldehyde condensation product, 100

Zinc
  dibenzyldithiocarbamate, 76
  dibutyldithiocarbamate, 75
  o,*o*-di-*n*-butyl phosphorodithioate, 94
  diethyldithiocarbamate, 46, 47, 75
  dimethyldithiocarbamate, 74
  ethyl phenyl dithiocarbamate, 76
  isopropyl xanthate, 93
  oxide, 59
  pentamethylene dithiocarbamate, 76
  salt of 2-mercaptobenzothiazole, 84
  stearate, 59

Subject Index

Accelerators, 72–95
Activators, 58–61
Acute toxicity
 definition, 134–5
 EEC grades, 148
 tests, 147–9
Adenoma, 136
Adsorber tube sampling, 168
Alcohols, 123–4
Aliphatic hydrocarbon solvents, 121
Ames test, 151–4
Angiosarcoma, 136
'Antabuse' effect, 91
Antidegradants, 99–114
Antioxidants, 99–114
Area sampling, 157–8
Aromatic amines in rubber chemicals, 29–34
Aromatic hydrocarbon solvents, 120–1
Aromatic oils, 35–44, 68
 carcinogenicity, 36
Asthma—occupational
 azodicarbonamide, 115
 phthalic anhydride, 98
Automatic dust monitors
 light scattering, 162–3
 piezoelectric, 162
 β-radiation, 161

Benzo[a]pyrene
 in carbon black, 37–9
 in mineral oil, 35–7

Benzo[a]pyrene—contd.
 in rubber factories, 39–44
 in urban air, 40, 42
Bladder cancer, 24–8
 in the European rubber industry, 27
 in the USA rubber industry, 27
Blowing agents, 115–19
British Rubber Manufacturers' Association/University of Birmingham epidemiological study, 4–13
Bubblers, sampling, 167–8

Carbon black
 carcinogenicity, 37–9, 57–8
 lung function effects, 56–7
 polycyclic aromatic hydrocarbon content, 37–8
Carboxyhaemoglobin production from exposure to methane chloride, 122
Carcinogenicity
 definition, 135–8
 IARC ratings, 152–3
 tests, 150–5
Cardiac effects of trichloroethylene, 122
Case/control studies, 4
Case, R.A.M., bladder cancer studies, 24–6
Cassidy and Wright versus ICI and Dunlop, 26
Ceiling limits, 157

Central nervous system damage
  dinitrosopentamethylenetetramine, 116
  phosphate esters, 70
Chlorinated solvents, 122
Chlorinated waxes, 69
Chromosome damage, tests for, 151–5
Chronic toxicity
  definition, 135
  tests, 149–55
Cigarette smoke
  bladder cancer, 31
  nitrosamines, 47
  polycyclic aromatic hydrocarbons, 42
Clastogens, 138
Co-carcinogens, 138
Colorimetric monitoring methods, 167–8
Control limits, 156
Corneal scarring—diethylthiourea, 87
Curing agents, 63–7
Cyclone, for respirable dust sampling, 159–60
Cytodiagnosis, 26
Continuous monitors
  dust, 161–3
  electrochemical, 173
  gas chromatography, 174
  infra-red, 172
  paper tape, 171
  solid-state, 173

De-phenylation of phenyl β-naphthylamine, 32–3, 102–3
Detector tube sampling, 166–7
Dose/response relationships, 145–7
Dust
  inhalation, 141–2
  measurement
    automatic, 161–3
    inhalable, 159
    respirable, 159–60
    total, 158–9
Dyestuffs manufacture—bladder cancer, 24–5

EEC toxicity grades, 148
Electrochemical monitors, 173
Epigenetic carcinogens, 138
Esters
  plasticisers, 70–1
  solvents, 124
Ethylene glycol
  monoalkyl ether esters, 124–7
  monoalkyl ethers, 124–7
Exposure limits, 156
Eye irritation
  definition, 134
  tests, 139

$FEV_{1 \cdot 0}$, 134
Fibrosis, 142
Fick's law of diffusion, 169
Fillers, 58–9
Filters, for dust sampling, 160
Foetotoxicity, 139
Forced Expiratory Volume, 134
Forced Vital Capacity, 134
Fume
  BRMA limit, 164
  composition, 164
  measurement, 163–5
FVC, 134

Gas chromatography
  in laboratory analysis of collected gases and vapours, 168
  portable monitors, 174
Gases—measurement, 166–74
Genotoxic carcinogens, 138
Glycol ether solvents, 124–7
Grab sampling, 166

Harvard School of Public Health, epidemiological studies, 19–21
Health and Safety Executive epidemiological study, 13–16
Healthy worker effect, 4
Hot rubber fume
  BRMA limit, 164

SUBJECT INDEX

Hot rubber fume—*contd.*
  composition, 164
  measurement, 164–5
Hydrocarbon solvents, 120–1
Hyperplasia, 135

Infra-red monitors, 172–3
Ingestion, 140
Inhalable dust,
  definition, 141–2
  measurement, 159
Inhalation, 140–3
Initiation—cancer, 136–7
Irritation
  eye, nose and throat, 134
  respiratory, 134
  skin, 133

'Jamaica Ginger' poisoning episode, 70

Keratitis—diethylthiourea, 87
Ketones, 123

Latent period—β-naphthylamine, 26
Lethal dose (50%), LD_{50}, 147
Leukaemia
  benzene, 120
  UK rubber industry, 5, 6, 11, 14
  USA rubber industry, 17–21
Lung cancer in the rubber industry
  Sweden, 21
  UK, 5–9, 11–12, 14, 15
  USA, 17, 18, 20
Lung function
  changes due to carbon black exposure, 56–7
  changes due to talc exposure, 58–9
  definition, 134

Magnusson and Kligman maximisation test, 148–9
MAK limits, 156
Mesothelioma, 136

Metabolic activation of β-naphthylamine, 33
Metabolic production of β-naphthylamine, 32–4
  102–3,107
Metastasis, 135
Micronucleus test, 154
Mineral fillers, 58–60
Monomers, 53–4
Mutagenicity, 138

Neurotoxicity
  dinitrosopentamethylenetetramine, 116
  n-hexane, 121
  methyl *n*-butyl ketone, 123
  tetramethylthiuram disulphide, 90
Nitrosamines
  concentrations in rubber production, 46–7
  dinitrosopentamethylenetetramine, 116
  mechanisms of production, 45–6
  N-nitrosodiphenylamine, 96

Oesophageal cancer, 11, 13, 14, 21
Oils, 68
Optic nerve damage—methanol, 124

Packed tube sampling, 168
Paper tape sampling, 171–2
Papilloma, 136
Passive samplers
  diffusion, 168–9
  permeation, 169–70
Peripheral neuropathy
  n-hexane, 121
  methyl *n*-butyl ketone, 123
Personal sampling, 157–8
Phagocytosis, 142
Plasticisers, 68–71
Pneumoconiosis
  carbon black, 56–7
  silica, 59–60
  sulphur, 63
  talc, 58–9

Polycyclic aromatic hydrocarbons
  in aromatic oils, 35–7
  in carbon black, 37–9
  in rubber factory atmospheres, 39–44
Polymer additives, 54–5
Polymers, natural and synthetic, 53–5
Primary skin irritation, tests, 148
Promotion, cancer, 138

Reinforcing agents, 56–61
Reproductive effects
  benzene sulphonyl hydrazide, 116
  N,N'-diphenyl-p-phenylenediamine, 106
  ethylene glycol mono alkyl ethers, 126–7
  thioureas, 85–9
Reproductive toxicity effects, 139
Resins, 69
Respirable dust
  definition, 141–3
  measurement, 159–60
Respiratory sensitisation, 134
  azodicarbonamide, 115
  phthalic anhydride, 98
Retarders, 96–8
Retrospective follow-up studies—definition, 3
Rubbers, natural and synthetic, 53–5

Sampling bag monitoring, 167
Sampling pumps, 159, 168
Sarcoma, 136
Sensitisation
  respiratory, 134
  skin, 133, 148–9
Sensitised paper tape sampling, 171–2
Short term cancer tests, 151–5
Skin
  absorption, 140
  irritation
    definition, 133
    tests, 148–9

Solid state monitors, 173
Solvents, 120–9
Spot sampling, 166
Stabilised flow sampling pumps, 159
Standardised mortality ratio, 3
Static sampling, 157–8
Stomach cancer in the rubber industry
  Switzerland, 21
  UK, 5, 6, 9–16
  USA, 17–21
Sub-acute toxicity tests, 149–50

Tackifying resins, 69
Teratogenesis
  benzene sulphonyl hydrazide, 116
  N,N'-diphenyl-p-phenylenediamine, 106
  ethylene glycol mono alkyl ethers, 127
  ethylene thiourea, 86
Teratogenicity, 139
Threshold limit values
  short term exposure limits, 157
  time weighted average, 156–7
Thyroid tumours—ethylene thiourea, 85–6
Total dust
  definition, 141
  measurement, 158–9
Toxicity
  acute, 134–5
  chronic, 135

University of North Carolina
  epidemiological studies, 16–19

Vapours
  inhalation, 143
  measurement, 166–74
Veys epidemiological studies, 16

Waxes, 68–9

NO LONGER THE PROPERTY
OF THE
UNIVERSITY OF R. I. LIBRARY